C++面向对象程序设计双语教程
（第3版）

刘嘉敏　马广焜　常　燕　编著
朱世铁　李树江

电子工业出版社
Publishing House of Electronics Industry
北京·BEIJING

内 容 简 介

本书在保持前两版特色的基础上，对部分章节内容进行了修改和补充。全书案例易懂、切合实际。本书共 8 章，围绕面向对象程序设计中类和对象的作用，介绍标准 C++中类与对象的定义和封装、继承、重载、多态、模板的概念及实现方法。

本书用通俗易懂的英语描述其内容，让初学者了解面向对象程序设计的原文表达；而且在各章节中的重要知识点和易混淆知识点处均有双语注解，有助于读者掌握面向对象的程序设计方法。

本书面向具有程序设计基础的读者，可作为高等院校计算机及相关专业的面向对象程序设计课程的双语教材。

未经许可，不得以任何方式复制或抄袭本书之部分或全部内容。
版权所有，侵权必究。

图书在版编目（CIP）数据

C++面向对象程序设计双语教程 / 刘嘉敏等编著. —3 版. —北京：电子工业出版社，2019.8
ISBN 978-7-121-36454-9

I. ①C… II. ①刘… III. ①C 语言－程序设计－双语教学－高等学校－教材 IV. ①TP312.8

中国版本图书馆 CIP 数据核字（2019）第 083125 号

责任编辑：赵玉山　　　特约编辑：张燕虹
印　　刷：北京捷迅佳彩印刷有限公司
装　　订：北京捷迅佳彩印刷有限公司
出版发行：电子工业出版社
　　　　　北京市海淀区万寿路 173 信箱　邮编：100036
开　　本：787×1 092　1/16　印张：17.25　字数：574 千字
版　　次：2013 年 2 月第 1 版
　　　　　2019 年 8 月第 3 版
印　　次：2019 年 8 月第 1 次印刷
定　　价：54.00 元

凡所购买电子工业出版社图书有缺损问题，请向购买书店调换。若书店售缺，请与本社发行部联系，联系及邮购电话：（010）88254888，88258888。
质量投诉请发邮件至 zlts@phei.com.cn，盗版侵权举报请发邮件至 dbqq@phei.com.cn。
本书咨询联系方式：（010）88254556，zhaoys@phei.com.cn。

前　　言

　　本书自第 1 版和第 2 版出版以来，在作者所在学校计算机科学与技术专业和其他高校相关专业教学中已使用 5 年，得到了广大师生的赞许，不失为高等院校相关专业课程的一本较好且实用性强的双语教材。

　　本书从面向对象程序设计的特点和工程应用角度出发，渐进式地组织各章节的知识点，采用既易懂又切合实际的实例，引导初学者进入面向对象程序设计之门。本书结合当今学生对新知识的认知程度，以保持原有特色为前提，在前两版的基础上，对各章内容进行了删减、增补和修改，适当地增加了案例和练习题，突出主线，使知识点的关联性更好，有助于初学者理解和掌握相关概念和实现方法。

　　本书在保持英文原汁原味的基础上，采用符合中国学生学习习惯且通俗易懂的英语描述，并针对重要的知识点，通过"教学目标""重点注释""句法""案例""概念加注框""思考题""词汇小贴士""案例学习""练习题"形式来组织内容，重点突出、易于学习。特别是本版对"重点注释""概念加注框"给出了相应的中文解释，而且采用黑体字突显其中的概念。此外，在正文中采用黑斜体字和斜体字突出了关键术语和代码中出现的变量和函数名，这样更便于学生阅读和掌握。

　　本书旨在培养学生掌握面向对象程序设计的基本概念、思想和方法，虽然本书内容是以标准 C++程序设计语言来描述的，但不是描述 C++的面向对象程序设计，其内涵与用其他语言描述面向对象程序设计是一致的，因而在许多概念上不过分强调 C++语言的细节，而是着重强调对面向对象程序设计的概念、思想和方法的认识。

　　面向对象程序设计是一门实践性较强的课程，因而上机实践是学习和巩固知识点必不可少的环节。为此，本书针对每个知识点配有完整的代码和运行结果，使学生可以确定程序预期结果，通过输出结果与程序语句联系在一起，为学生提供实践和自学的方式。书中所有源代码在 Visual Studio 2013 环境中调试和运行。

　　本书适合 40~54 学时教学，配有 PPT 教学课件，并提供实验指导书、实验题目详解代码、面向对象程序设计的课程设计题目，有需要的任课教师请垂询 jmliu@sut.edu.com。

　　刘嘉敏老师修订了第 1、3、4、5、8 章，马广焜老师修订了第 6、7 章并参与了第 8 章的修订，姚凯老师修订了第 2 章，常燕老师参与了第 3、4 章和练习题的修订，朱世铁老师参与了练习题和实验指导书的修订，李树江老师参与了第 6、7 章的修订。本书得到了沈阳工业大学信息科学与工程学院老师和学生、英国大学同行专家、美国南加州大学王博涵同学的帮助，以及作者家人的支持，而且书中引用了其他同行的工作成果，在此表示衷心感谢。

由于作者的英语水平有限，书中难免存在错漏和不妥之处，恳请读者提出宝贵意见。

编　著　者

Contents

Chapter 1 Introduction ··· 1
  1.1 Overview of Programming ··· 1
      1.1.1 What Is Programming? ·· 1
      1.1.2 How to Write a Program? ·· 3
  1.2 Programming Methodologies ·· 5
      1.2.1 Structured Programming ·· 5
      1.2.2 Object-Oriented Programming ······································ 8
  1.3 Characteristics of Object-Oriented Programming ··················· 10
  1.4 C++ Programming Language ·· 13
      1.4.1 History of C and C++ ·· 13
      1.4.2 Learning C++ ··· 15
  Word Tips ·· 16
  Exercises ·· 17

Chapter 2 Basic Facilities — *Shifting from C to C++ Programs* ········ 18
  2.1 C++ Program Structure ·· 18
  2.2 Input/Output Streams ·· 21
      2.2.1 Input Stream ··· 21
      2.2.2 Output Stream ·· 22
  2.3 Constants ·· 23
  2.4 Functions ·· 25
      2.4.1 Function Declarations ·· 25
      2.4.2 Function Definitions ·· 26
      2.4.3 Default Arguments ·· 28
      2.4.4 Inline Functions ·· 30
      2.4.5 Function Overloading ·· 30
  2.5 References ·· 35
      2.5.1 Reference Definition ··· 35
      2.5.2 Reference Variables as Parameters ······························ 39
      2.5.3 References as Returning Values ·································· 40
      2.5.4 Reference as Left-Hand Values ·································· 42
  2.6 Namespaces ·· 43
  Word Tips ·· 47

Exercises ·· 48

Chapter 3 Foundation of Classes and Objects—*Data Abstraction and Definition of Classes* ·· 52

3.1   Introduction to Structures ··· 52
    3.1.1   Defining a Structure in C++ ··· 52
    3.1.2   Accessing Members of Structures ·· 53
    3.1.3   Structures with Member Functions ·· 55

3.2   Data Abstraction and Classes ·· 56
    3.2.1   Data Abstraction ··· 56
    3.2.2   Defining Classes ·· 57
    3.2.3   Defining Objects ·· 58
    3.2.4   Accessing Member Functions ··· 59
    3.2.5   In-Class Member Function Definition ·· 61
    3.2.6   File Structure of an Abstract Data Type ·· 63

3.3   Information Hiding ·· 65

3.4   Access Control ··· 66

3.5   Constructors ··· 69
    3.5.1   Definition of Constructors ·· 69
    3.5.2   Overloading Constructors ·· 70
    3.5.3   Constructors with Default Parameters ·· 71

3.6   Destructors ·· 74
    3.6.1   Definition of Destructors ··· 74
    3.6.2   UML Diagram for Classes ·· 75
    3.6.3   The Order of Constructor and Destructor Calls ···································· 76

3.7   Encapsulation ··· 78

3.8   Case Study: A GradeBook Class ··· 80

Word Tips ··· 82

Exercises ·· 83

Chapter 4 Advance of Classes and Objects —*Further Definition of Class Members and Objects* ··· 87

4.1   Constant Member Functions and Constant Objects ·· 87

4.2   this[①] Pointers ··· 89

4.3   Static Members ··· 91

---

① 注：因是特殊的指针的关键字，故首字母非大写。

  4.3.1 Static Data Members ·· 93
  4.3.2 Static Member Functions ··· 96
4.4 Free Store ··· 97
4.5 Object Members ·· 101
  4.5.1 Definition of Object Members ································· 101
  4.5.2 The Order of Constructors and Destructors for Member Objects ········ 105
  4.5.3 Object Members with Default Constructors ················ 105
  4.5.4 Class Members by Using Initializers ··························· 106
4.6 Copy Members ··· 107
  4.6.1 Definition of Copy Constructors ································ 108
  4.6.2 Shallow Copy and Deep Copy ··································· 110
4.7 Arrays of Objects ·· 118
  4.7.1 Initialize an Object Array by Using a Default Constructor ·············· 118
  4.7.2 Initialize an Object Array by Using Constructors with Parameters ······ 121
4.8 Friends ··· 122
  4.8.1 Friend Functions ··· 122
  4.8.2 Friend Classes ·· 125
4.9 Case Study: Advance of the GradeBook Class ························· 126
Word Tips ··· 132
Exercises ··· 132

Chapter 5 Operator Overloading ··· 137

5.1 Introduction to Operator Overloading ··································· 137
5.2 Operator Functions ·· 138
  5.2.1 Overloaded Operators ··· 138
  5.2.2 Operator Functions ·· 138
5.3 Binary and Unary Operators ··· 142
  5.3.1 Overloading Binary Operators ··································· 142
  5.3.2 Overloading Unary Operators ··································· 143
5.4 Overloading Combinatorial Operators ·································· 147
5.5 Mixed Arithmetic of User-Defined Types ····························· 151
5.6 Type Conversion of User-Defined Types ····························· 152
  5.6.1 Converting a Built-In Type to a User-Defined Type ······· 152
  5.6.2 Converting a User-Defined Type to a Built-In Type ······ 153
5.7 Case Study: A MyInteger Class ··· 155
Word Tips ··· 160
Exercises ··· 160

Chapter 6   Inheritance ········ 163
6.1   Class Hierarchies ········ 163
6.2   Derived Classes ········ 164
6.2.1   Declaration of Derived Classes ········ 164
6.2.2   Structure of Derived Classes ········ 165
6.3   Constructors and Destructors of Derived Classes ········ 168
6.3.1   Constructors of Derived Classes ········ 168
6.3.2   Destructors of Derived Classes ········ 171
6.3.3   The Calling Order of Derived Class Objects ········ 172
6.3.4   Inheritance and Composition ········ 175
6.4   Member Functions of Derived Classes ········ 175
6.4.1   Defining a Member Function ········ 175
6.4.2   Overriding Member Functions ········ 177
6.5   Access Control ········ 179
6.5.1   Access Control in Classes ········ 179
6.5.2   Access to Base Classes ········ 180
6.6   Multiple Inheritance ········ 184
6.6.1   Declaration of Multiple Inheritance ········ 185
6.6.2   Constructors of Multiple Inheritance ········ 187
6.7   Virtual Inheritance ········ 188
6.7.1   Multiple Inheritance Ambiguities ········ 188
6.7.2   Trying to Solve Inheritance Ambiguities ········ 189
6.7.3   Virtual Base Classes ········ 191
6.7.4   Constructing Objects of Multiple Inheritance ········ 194
6.8   Case Study: The iWatch Class ········ 195
Word Tips ········ 201
Exercises ········ 202

Chapter 7   Polymorphism and Virtual Functions ········ 212
7.1   Polymorphism ········ 212
7.1.1   Introduction to Polymorphism ········ 212
7.1.2   Binding ········ 213
7.2   Virtual Functions ········ 216
7.2.1   Definition of Virtual Functions ········ 216
7.2.2   Extensibility ········ 219
7.2.3   Principle of Virtual Functions ········ 221
7.2.4   Virtual Destructors ········ 223

  7.2.5 Function Overloading and Function Overriding ········· 224
 7.3 Abstract Base Classes ························· 227
 7.4 Case Study: A Mini System ······················· 230
 Word Tips ··································· 235
 Exercises ··································· 235

Chapter 8 Templates ································ 241
 8.1 Introduction to Templates ······················· 241
 8.2 Function Templates ·························· 242
  8.2.1 Definition of Function Templates ················· 242
  8.2.2 Function Template Instantiation ·················· 244
  8.2.3 Function Template with Different Parameter Types ········· 246
  8.2.4 Function Template Overloading ·················· 247
 8.3 Class Templates ···························· 249
  8.3.1 Definition of Class Templates ··················· 249
  8.3.2 Class Template Instantiation ···················· 251
 8.4 Non-Type Parameters for Templates ·················· 253
 8.5 Derivation and Class Templates ···················· 255
 8.6 Case Study: A Vector Class Template ················· 257
 Word Tips ··································· 262
 Exercises ··································· 263

References ···································· 264

Chapter 1 Introduction

High thoughts must have high language.
—Aristophanes

Objectives

- To understand what a computer programming is
- To be able to list the basic phases involved in writing a computer program
- To recognize two programming methodologies
- To know about the characteristics of object-oriented programming

1.1 Overview of Programming

1.1.1 What Is Programming?

Much of human behavior and thought is characterized by logical sequences. Since infancy, you have been learning how to behave and how to perform tasks. And you have learned to expect certain behavior from other people.

On the broad scale, mathematics could never have been developed without logically sequenced steps for solving problems and proving theorems. Mass production would never have worked without operations taking place in a certain order. Our whole civilization is based on the order of things and actions. We create order, both consciously and unconsciously, through a process that we call programming.

Programming is planning how to solve a problem. No matter what method is used—the pencil and paper, slide rule, adding a machine, or computer—problem solving requires programming. Of course, how one program depends on the device one uses in problem-solving.

Back to programming—"planning how to solve a problem", note that we are not actually solving a problem. The computer is going to do that for us. If we could solve the problem by ourselves, we would have no need to write the program. The premise for a program is that we don't have the time, tenacity or memory capabilities to solve the problem, but we do know how to solve it, therefore we can instruct a computer to do it for us.

> **Programming** is planning the performance of a task or an event.
>
> **Computer programming** is the process of planning a sequence of steps for a computer to follow.
>
> **Computer program** is a sequence of instructions to be performed by a computer.
>
> 程序设计是规划一个任务或一个事件的执行过程。
>
> 计算机程序设计是设计一系列的步骤让计算机来执行的过程。
>
> 计算机程序是由计算机执行的一系列指令。

A simple example of this is calculating what the sum of all integers from 1 to 10,000 is. If you wanted to, you could sit down with a pencil and paper or a calculator and work this out however the time needed, plus the likelihood that at some point you might make a mistake, rendering that an undesirable option. Instead, you can write and run a program to calculate this sum in less than 5 minutes.

Example 1-1: An example of programming.

```
/*-------------------------------------------------------------------
 * File: example1_1.c
 * The program calculates the sum of all integers from 1 to 10,000.
 *-------------------------------------------------------------------*/
1   #include <stdio.h>
2   #define MAX 10000
3
4   int main()
5   {
6       long sum = 0, number;
7       for( number = 1; number <= MAX; number ++)
8       {
9           sum += number;
10      }
11      printf("The sum of all integers from 1 to %ld is : %ld\n", MAX, sum);
12      return 0;
13  }
```

This example gives the result of 50,005,000. Meanwhile, you can verify this, as you know that the sum of integers from 1 to N can be calculated as

$(N+1) \times (N/2)$

$(10000 + 1) \times (10000/2) = 10001 \times 5000 = 50005000$

So, you have solved the problem of how to calculate the sum of all integers from 1 to 10,000 and the computer has solved the problem of calculating the sum of all integers from 1 to 10,000.

The computer allows us to do tasks more efficiently, quickly, and accurately than we could by hand—if we could do by hand at all. In order to use this powerful tool, we must specify what we want to do and the order in which we want to be done. We could do this through programming.

However, we have to know the major differences between human and computers. Human has the judgment and free will and will not run any instruction they deem not required or nonsensical, whereas a computer will do exactly what it is told with no judgment on the need or sanity of the instruction.

1.1.2 How to Write a Program?

To write a sequence of instructions for a computer to follow, we must go through a three-phase process: problem-solving, implementation, and maintenance (see Figure 1-1).

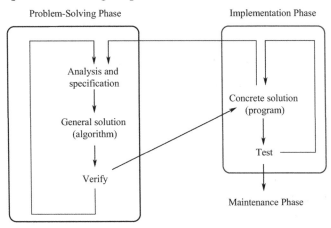

Figure 1-1 Programming process

Phase 1: Problem-Solving

(1) *Analysis and specification*. Understand (define) the problem and what the solution must do.

Every program starts with the specification. This may be several hundred-page documents from your latest client or one small paragraph from your professor and pretty much anything in-between.

The specification is very important, and the specification writing is a whole sub-branch of programming. The important thing is that the specification is correct, otherwise, to use a well-known computing adage, garbage in garbage out.

However, the specifications are rarely perfect, and it is not at all uncommon for the specification to go through several iterations before being finally agreed on.

(2) *General solution (algorithm)*. Develop a logical sequence of steps that solves the problem.

Having found out how to solve the problem you can then jump in and start coding. Right? Wrong! It is the time to do some program design, now. How much design is required, and where the design is stored rather depend on the complexity of the program, the experience of the user, and the purpose of the program. For instance, a simple program being written by an experienced programmer just as a temporary project tool (i.e. will only be in use for a day or two days) probably only requires a little thought about the program design.

The design, once you have it, is not set in stone. It just gives the current ideas about how the program will be written. Normally, the final code of a project does not match the initial design exactly. The design will be also changed as the specification changes. However, the specification will give a clear start point for coding and will ultimately lead to a better and more maintainable code.

(3) *Verify.* Follow the steps exactly to see if the solution does solve the problem.

Phase 2: Implementation

(1) *Concrete solution (program).* Translate the algorithm into a programming language.

(2) *Test.* Follow the instructions to check the results manually. If you find errors, analyze the program and algorithm to determine the source of the errors, and then make corrections.

Testing does not have to involve running any code. Source review by your peers is a good form of testing too. You sit down around a table and go through the code line by line and examine it for logic errors, conformance to standards and programming errors. This can actually throw up errors that are not made obvious from testing the software by running it.

Once a program has been written, it enters the third phase: maintenance.

Phase 3: Maintenance

(1) *Use.* Use the program.

(2) *Maintain.* Modify the program to meet the changing requirements or correct any errors that showed up in using it.

The programmer begins the programming process by analyzing the problem and developing a general solution called an ***algorithm***. Understanding and analyzing a problem take up much more time than what Figure 1-1 implies. They are the core of the programming process.

> **Algorithm** is a step-by-step procedure for solving a problem in a finite amount of time.
> 算法是在有限的时间内逐步解决问题的过程。

An algorithm is a verbal or written description of a logical sequence of actions (or events). After developing a general solution, the programmer tests the algorithm, walking through each step mentally or manually. If the algorithm does not work, the programmer repeats the problem-solving process, analyzes the problem again and comes up with another algorithm.

When the programmer is satisfied with the algorithm, he/she translates it into a programming language. Although a programming language is simple in form, it is not always easy to use. Programming forces you to write very simple, exact instruction. Translating an algorithm into a programming language is called *coding* the algorithm.

Once a program has been put into use, it is often necessary to modify it. Modification may involve fixing an error that is discovered during the use of the program or changing the program in response to changes in the user's requirements. Each time the program is modified, it is necessary to repeat the problem-solving and implementation phases for the modified aspects of the program. This phase of the programming process is known as *maintenance* and actually accounts for the majority of the effort expended on most programs.

In a word, the problem-solving, implementation and maintenance phases constitute the *program's life cycle*.

1.2 Programming Methodologies

Programming methodology deals with the analysis, design and implementation of programs. There are many forms of programming methodology. The purpose of making explicit awareness of programming methodology is to allow the programmer to be aware of the processes and procedures which they use when constructing programs.

Data Abstraction

One of the keys to successful programming is the concept of abstraction. Abstraction is crucial to building the complex software systems. A good definition of abstraction comes from and can be summed up as concentrating on the aspects relevant to the problem and ignoring those that are not important at the moment.

The psychological notion of abstraction allows one to concentrate on a problem at a certain level of generalization regardless of the irrelevant low-level details; use of abstraction also allows one to work in an environment with familiar concepts and terms without the need to transform it to an unfamiliar structure.

There are two top approaches to programming design, that is, the structured approach and the object-oriented approach, which are outlined below.

1.2.1 Structured Programming

Dividing a problem into smaller sub-problems is called *structured design*. Each sub-problem is then analyzed and a solution is obtained to solve the sub-problem. The solutions to all sub-problems are combined to solve the overall problem. This process of implementing a structured design is called *structured programming*. The *structured-design* approach is also known as the top-down design, stepwise refinement and modular programming (see Figure 1-2).

For using structured programming, there are many good reasons: codes are easier to understand and errors are easier to find. An error will always be localized to a subroutine or function rather than buried somewhere in a mass of code. The scope of variables can be controlled more easily. With the reuse of codes—as well as being reused within a single application, the modular programming allows codes to be used in multiple applications. Thus, programs are easier to design—the designer just needs to think about the high-level functions. Collaborative programming is possible—modular programming enables more than one programmer to work on a single application at the same time while the codes can be stored across multiple files.

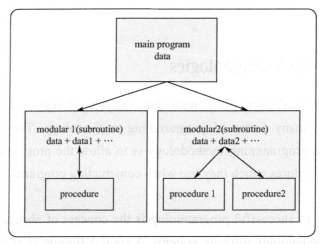

Figure 1-2   Structured programming (or Modular programming)

Now, we write a simple program which calculates the circumferences and areas of different sized circles. We design it in two ways.

Example 1-2: A simple example without structured programming.

```
/*-----------------------------------------------------------------
* File: example1_2.c
* The program calculates the circumferences and areas of different sized circles without using
* structured programming.
*-----------------------------------------------------------*/
1     #include <stdio.h>
2     #define PI 3.14156926
3     int main()
4     {
5         float r1 = 1.0;
6         float c1 = 2 * PI * r1;
7         float a1 = PI * r1 * r1;
8         printf("The circumference is %6.2f, the area is %6.2f", c1, a1);
9
10        float r2 = 2.0;
```

```
11      float c2 = 2 * PI * r2;
12      float a2 = PI * r2 * r2;
13      printf("The circumference is %6.2f, the area is %6.2f", c2, a2);
14
15      float r3 = 3.0;
16      float c3 = 2 * PI * r3;
17      float a3 = PI * r3 * r3;
18      printf("The circumference is %6.2f, the area is %6.2f", c3, a3);
14
15      float r4 = 4.0;
16      float c4 = 2 * PI * r4;
17      float a4 = PI * r4 * r4;
18      printf("The circumference is %6.2f, the area is %6.2f", c4, a4);
19
20      return 0;
21   }
```

Here the same functionality is repeated and work indeed. However, the programmer can work more efficiently by using structured programming.

The aim of the programmer is now to ensure that the code can be reused as much as possible. Such goal can be achieved through by identifying the code that can be placed in separate functions and subroutines.

Example 1-3: A simple example with structured programming.

```
/*----------------------------------------------------------------------
 * File: example1_3.c
 * The program calculates the circumferences and areas of different sized circles with using
 * structured programming.
 *----------------------------------------------------------------------*/
1    #include <stdio.h>
2    #define PI 3.14156926
3
4    float circum (float r)
5    {    return 2 * PI * r;    }
6    float area (float r)
7    {    return PI * r * r;    }
8    void print (float r)
9    {    printf("The circumference is %6.2f, the area is %6.2f", circum(r), area(r));    }
10
11   int main()
12   {
13       print(1.0);
14       print(2.0);
15       print(3.0);
16       print(4.0);
17       return 0;
```

The output is the same as above, but this time the chance of the script operation being altered due to a typing error is greatly reduced. In addition, if there is an error, the task of correcting it is made much simpler. The programmer can also use the functionality throughout their application without worrying about having to rewrite the code, thereby improving the code and the time efficiency.

> **Structured programming** is a programming paradigm that enforces a logical structure on the program being written to make it more efficient and easier to understand and modify.
>
> It is a kind of top-down design, stepwise refinement and modular programming. It divides a problem into several sub-problems, and each sub-problem is then analyzed, afterward, a solution is obtained to solve the sub-problem. The solutions to all sub-problems are finally combined to solve the overall problem.
>
> 结构化程序设计是一种编程范式,它使得编写的程序更有逻辑结构,以便让它更有效、更容易理解和修改。
>
> 它是一种自顶向下、逐步精细、模块化的程序设计方法。它将一个问题被划分为若干子问题,对每个子问题进行分析和求解。最终所有子问题的解构成这个问题的解。

1.2.2 Object-Oriented Programming

Object-Oriented Programming (**OOP**) is a widely used programming methodology. In OOP, the first step is to identify the components called ***objects***, which form the basis of the solution, and to determine how these objects interact with one another, as shown in Figure 1-3. The next step is to specify for each object the relevant data and possible operations to be performed on the data.

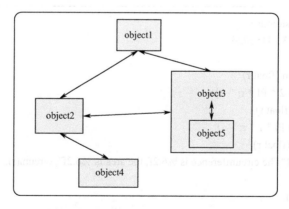

Figure 1-3 Object-oriented programming

Object-oriented programming is centered on the object. An object is a programming element that contains both the data and the procedures that can be operated on the data. The objects contain, within themselves, both the information and the ability to manipulate the information.

Let us change the program in Example 1-3 using object-oriented programming.

Example 1-4: A simple example with OOP.

```
/*--------------------------------------------------------------------
 * File: example1_4.cpp
 * The program calculates the circumferences and areas of different sized circles with using OOP.
 *--------------------------------------------------------------------*/
1    #include <stdio.h>
2    #define PI 3.14156926
3
4    class Circle{
5    public:
6       Circle(float R)
7       {   r = R;   }
8       float circum()
9       {   return 2 * PI * r;   }
10      float area()
11      {   return PI * r * r;   }
12      void print()
13      {   printf("The circumference is %6.2f, the area is %6.2f", circum(), area());   }
14   private:
15      float r;
16   };
17
18   int main()
19   {
20      Circle c1(1.0), c2(2.0), c3(3.0), c4(4.0);
21      c1.print();
22      c2.print();
23      c3.print();
24      c4.print();
25      return 0;
26   }
```

From the first glance in Example 1-4, you may see that everything following "public:" is the functions: *circum*, *area* and *print*, while "private:" denotes the *r* variable. This is the definition of an object, called a ***class***. We can see that class *Circle* combines data and operations on data into a single unit. In the *main* function, we create four different objects of class *Circle*, i.e. *c*1, *c*2, *c*3 and *c*4, and deal with them directly.

The primary differences between the two approaches are their use of data. In a structured program, the design centers on the rules or procedures for processing the data. The procedures, implemented as functions in C++, are the focus of the design. The data objects are passed to the functions as parameters. The key question is how the functions will transform the data they receive to either storage or further processing. In an object-oriented program, the design centers on the objects that contain the data and the necessary functions to process the data. Procedural programming has been the mainstay of computer science since its beginning. However, object-oriented programming is used as a mainstream program method today.

Think it Over

Why use classes in OOP instead of functions?

1.3 Characteristics of Object-Oriented Programming

Object-oriented programming (OOP) is a programming paradigm that uses "objects"—data structures consisting of data fields and methods together with their interactions—to design applications and computer programs. Programming techniques may include features such as data abstraction, encapsulation, polymorphism, and inheritance. It was not commonly used in the mainstream software application development until the early 1990s. Many modern programming languages now support OOP.

The root of object-oriented programming can be traced to the 1960s. As hardware and software became increasingly complex, manageability often became a concern. Researchers studied ways to maintain software quality and developed object-oriented programming in part to address common problems by strongly emphasizing discrete, reusable units of programming logic. Such technology focuses on the data rather than the processes, with programs composed of self-sufficient modules ("classes"), each instance of which ("objects") contains all the information needed to manipulate its own data structure ("members"). This is in contrast to the existing modular programming which had been dominant for many years that focused on the *function* of a module, rather than specifically the data. Nevertheless, it equally provided for code reuse, and self-sufficient reusable units of programming logic, enabling collaboration through the use of linked modules (subroutines).

An object-oriented program may thus be viewed as a collection of interacting *objects*, as opposed to the conventional model, in which a program is seen as a list of tasks (subroutines) to perform. In OOP, each object is capable of receiving messages, processing data, and sending messages to other objects. Each object can be viewed as an independent "machine" with a distinct role or responsibility.

Class

A *class* is a user-defined data type which contains variables, properties and methods in it. A class defines the abstract characteristics of a thing (object), including its characteristics (its attributes, fields or properties) and the thing's behaviors (the things it can do, or methods, operations or features). One might say that a class is a *blueprint* or *factory* that describes the nature of something.

For example, the *Dog* class would consist of traits shared by all dogs, such as breed and fur color (characteristics), and the ability to bark and sit (behaviors). Classes provide modularity and structure in an object-oriented computer program. A class should typically be recognizable by a non-programmer familiar with the problem domain, meaning that the characteristics of the class should make sense in context. Also, the code for a class should be relatively self-contained (generally using encapsulation). Collectively, the properties and methods defined by a class are called members.

Method

The ***method*** is a set of procedural statements for achieving the desired result. It performs different kinds of operations on different data types. In a programming language, methods (sometimes referred to as "functions") are verbs. For example, *Lassie*, being a *Dog* object, has the ability to bark. So *bark* is one of *Lassie*'s methods. She may have other methods as well, for example, *sit* or *eat* or *walk* or *save*. Within the program, using a method usually affects only one particular object; all dogs can bark, but you need only one particular dog to do the barking.

Message Passing

Message passing is the process by which an object sends data to another object or asks the other object to invoke a method which is known to some programming languages as interfacing. For example, the object called *Breeder* may tell the *Lassie* object to sit by passing a "*sit*" message which invokes *Lassie*'s "*sit*" method.

Abstraction

Abstraction is one of the most powerful and vital features provided by object-oriented programming. The concept of abstraction relates to the idea of hiding data that are not needed for the presentation. The main idea behind data abstraction is to clearly separate the properties of data type and the associated implementation details. This separation is achieved so that the properties of the abstract data type are visible to the user interface while the implementation details are hidden. Thus, abstraction forms the basic platform for the creation of user-defined data types called objects. ***Data abstraction*** is the process of refining data to its essential form. An ***Abstract Data Type*** (ADT) is a data type that is defined in terms of the operations that it supports but not in terms of its structure or implementation.

In object-oriented programming language C++, it is possible to create and provide an interface that accesses only certain elements of data types. The programmer can decide which user to give or grant access to and hide the other details. This concept is called data hiding which is similar in concept to data abstraction.

For example, in a touch screen at the railway station or ATM machine in the bank, we just use the touch screen application to satisfy our needs, however, we don't see what is happening inside its software or about its OS.

Encapsulation

Encapsulation conceals the functional details of a class from objects that send messages to it.

For example, the *Dog* class has a *bark* method. The code for the *bark* method defines exactly how a bark happens (e.g., by an *inhale* method and then an *exhale* method, at a particular pitch and volume). *Timmy*, *Lassie*'s friend, however, does not need to know exactly how she barks. Encapsulation is achieved by specifying which classes may use the members of an object. The result is that each object exposes to any class a certain *interface*—those members accessible to that class. The reason for encapsulation is to prevent clients of an interface from depending on those parts of the implementation that are likely to change in the future, thereby allowing those changes to be made more easily, that is, without changes to clients. For example, an interface can ensure that puppies can only be added to an object of the *Dog* class by code in that class. Members are often specified as public, protected or private, determining whether they are available to all classes, sub-classes or only the defining class.

Inheritance

Inheritance is a process in which a class inherits all the state and behavior of another class. This type of relationship is called child-parent or is a relationship. "Subclasses" are more specialized versions of a class, which inherit attributes and behaviors from their parent classes, and can introduce their own.

For example, the *Dog* class might have sub-classes called *Collie*, *Chihuahua*, and *GoldenRetriever*. In this case, *Lassie* would be an instance of the Collie subclass. Suppose the *Dog* class defines a method called *bark* and a property called *furColor*. Each of its sub-classes (*Collie*, *Chihuahua*, and *GoldenRetriever*) will inherit these members, meaning that the programmer only needs to write the code for them once.

Each subclass can alter its inherited traits. For example, the *Collie* subclass might specify that the default *furColor* for a *Collie* is brown-and-white. The *Chihuahua* subclass might specify that the *bark* method produces a high pitch by default. Subclasses can also add new members. The *Chihuahua* subclass could add a method called *tremble*. Thus, an individual *Chihuahua* instance would use a high-pitched *bark* from the *Chihuahua* subclass, which in turn inherited the usual *bark* from *Dog*. The *Chihuahua* object would also have the *tremble* method, but *Lassie* would not, because she is a *Collie*, not a *Chihuahua*. In fact, inheritance is an "a...is a" relationship between classes, while instantiation is an "is a" relationship between an object and a class: a *Collie* is a *Dog* ("a... is a"), but *Lassie* is a *Collie* ("is a"). Thus, the object named *Lassie* has the methods from both classes *Collie* and *Dog*.

Polymorphism

Polymorphism allows the programmer to treat derived class members just like their parent class's members. More precisely, Polymorphism in object-oriented programming is the ability of objects belonging to different data types to respond to calls of methods of the same name, each one according to an appropriate type-specific behavior. One method, or an operator such as +, -, or *, can be abstractly applied in many different situations. If a *Dog* is commanded to a *speak* method, this may elicit a *bark* method. However, if a *Pig* is commanded to *speak* this may elicit an *oink* method. Each subclass overrides the *speak* method inherited from the parent class *Animal*.

1.4 C++ Programming Language

"As long as there were no machines, programming was no problem at all; when we had a few weak computers, programming became a mild problem and now that we have gigantic computers, programming has become an equally gigantic problem. In this sense the electronic industry has not solved a single problem, it has only created them—it has created the problem of using its product" (E.W. Dijkstra, Turing Award Lecture, 1972).

Nowadays, there are many programming languages, for example, C, C++, Ada, Pascal, Prolog, FORTRAN, Modula3, Lisp, Java, and Scheme. This alphabet soup is the secret power of modern software engineering. As high-level computer programming languages, they provide enormous flexibility and abstraction. Programmers are separated from the physical machine, allowing them to create complex problem solutions without fretting with the troubles of ones and zeros. This idea is clarified by an analogy. If we had to think about every phonetic sound made while speaking, communication of abstract ideas would be nearly impossible. Much the same way, programming directly with ones and zeros would focus the designer's attention on trivial hardware details instead of on designing abstract solutions. Considering the historical trend that created high-level programming, we believe that certain reasonable predictions can be made regarding future advances.

1.4.1 History of C and C++

C evolved from two previous programming language BCPL and B. BCPL was developed in 1967 by Martin Richards as a language for writing operating systems software and compilers. Ken Thompson modeled many features in his language B after their counterparts in BCPL and used B to create early versions of the UNIX operating system at the Bell

Laboratories in 1970 on a DEC PDP-7 computer. Both BCPL and B were "typeless" language—every data item occupied one "word" in memory while the burden of treating a data item as a whole number or a real number, for example, fell on the shoulders of the programmer.

The C language was evolved from B by Dennis Ritchie at Bell Laboratories and was originally implemented on a DEC PDP_11 computer in 1972. C uses many important concepts of BCPL and B while adding the data typing and other features. C initially became widely known as the development language of the UNIX operating system. Today, most operating systems are written in C and/or C++.

By the late 1970s, C has evolved into what is now referred to as "traditional C", "classic C", or "Kernighan and Ritchie C". The widespread use of C with various types of computers (sometimes called hardware platforms) unfortunately led to many variations. There were similar, but often incompatible. This was a serious problem for program developers who needed to write portable programs that would run on multiple platforms. It became clear that a standard version of C was needed. In 1983, the X3J11 technical committee was created under the American National Standard Committee on Computers and Information Processing (X3) to "provide an unambiguous and machine-independent definition of the language". In 1989, the standard was approved. ANSI cooperated with the International Standards Organization (ISO) to standardize C worldwide, when the joint standard document was published in 1990 and referred to as the ANSI/ISO 9899:1990. Copies of this document may be ordered from ANSI. As the second edition of the one published by Kernighan and Ritchie in 1988, reflects this version is called ANSI C, a version of the C language now used worldwide.

C++, an extension of C, was developed by Bjarne Stroustrup at the Bell Laboratories during 1983-1985. Prior to 1983, Bjarne Stroustrup added features to C and formed what he called "C with classes". He had combined the use of classes and object-oriented features of Simula, having the power and efficiency of C. The term C++ was first used in 1983.

The name C++ was coined by Rick Mascitti in the summer of 1983. The name signifies the evolutionary nature of the changes from C; "++" is the C increment operator. The slightly shorter name "C+" is a syntax error; it has also been used as the name of an unrelated language. Connoisseurs of C semantics find C++ inferior to ++C. The language is not called D as it is an extension of C, and does not attempt to remedy problems by removing features.

Like C, ISO also published the first international standard for C++ in 1998, known as C++98. C++98 includes the Standard Template Library (STL), the conceptual development of which began in 1979. In 2003 and 2005, ISO revised problems in C++98 respectively. A new C++ standard (known as C++11) was approved by ISO in 2011. C++ 11 has added new features into the core language and the standard library. These new features are very useful for advanced C++ programming.

1.4.2 Learning C++

As Bjarne Stroustrup mentioned, C++ is a general-purpose programming language with advantages towards the system programming that

- is a better C,
- supports data abstraction,
- supports object-oriented programming, and
- supports generic programming.

The most important thing to do when learning C++ is to focus on concepts and not get lost in language-technical details. The purpose of learning a programming language is to become a better programmer, that is, to become more effective at designing and implementing new systems and at maintaining old ones. For this, the appreciation of programming and design techniques is far more important than the understanding of details, which comes with time and practice.

C++ supports a variety of programming styles. All are based on strong static type checking, and most aim at achieving a high level of abstraction and a direct representation of the programmer's ideas. Each style can achieve its aims effectively while maintaining run-time and space efficiency. A programmer originally using a different language (say C, Fortran, Smalltalk, Pascal, or Modula-2) should realize that to gain the benefits of C++, they must spend time learning and internalizing programming styles and techniques suitable to C++. The same situation applies to programmers used to an earlier and less expressive version of C++.

Thoughtlessly applying techniques effective in one language to another typically leads to awkward, poorly performing, and hard-to-maintain code. Such code is also most frustrating to write because every line of code and every compiler error message reminds the programmer that the language used differs from "the old language". You can write in the style of Fortran, C, Smalltalk, etc., in any language, but doing so is neither pleasant nor economical in a language with a different philosophy. Every language can be a fertile source of ideas on how to write C++ programs. However, ideas must be transformed into something that fits with the general structure and type system of C++ in order to be effective in different contexts.

C++ supports a gradual approach to learning. How you approach learning a new programming language depends on what you already know and what you aim to learn. There is no single approach that suits everyone. My assumption is that you are learning C++ to become a better programmer and designer. Ergo, your purpose in learning C++ is not simply to learn a new syntax for doing things the way you used to, but to learn new and better ways of building systems. This has to be done gradually because acquiring any significant new skill takes time and requires practice. Consider how long it would take to learn a new natural language well or to learn to play a new musical instrument well. Becoming a better system designer is easier and faster, but not as much easier and faster as most people would like it to be.

Word Tips

abstract *adj.* 抽象的	implement *vt.* 执行，实现
abstraction *n.* 抽象，抽取	individual *adj.* 个别的，单独的
accurate *adj.* 精确的	infancy *n.* 早期
actually *adv.* 实际上	instruction *n.* 命令，指示
adage *n.* 言语，格言	integer *n.* 整数
algorithm *n.* 运算法则	internal *adj.* 内部的
analogy *n.* 类似，相似，类推	internalize *v.* 吸收同化
analysis *n.* 分析，分析报告	invoke *vt.* 调用
broad *adj.* 宽的，广的	iteration *n.* 反复，迭代
calculate *vi./vt.* 计算，估计	logical *adj.* 逻辑上的
chihuahua *n.* 吉娃娃狗	mainstay *n.* 支柱，骨干
clarify *v.* 使清楚	maintain *vt.* 维护
client *n.* 委托人，顾客	methodology *n.* 方法
code *n.* 代码	modular *adj.* 模块的
cohesion *n.* 聚合	module *n.* 单元，单位
concrete *adj.* 实体的，有形的	multiple *adj.* 多重的，多种多样的
coupling *n.* 耦合	paradigm *n.* 范式
crux *n.* 难点，关键	parameter *n.* 参数
deem *vi./vt.* 认为，相信	phase *n.* 阶段，时期
detail *n.* 细节，小事	phonetic *adj.* 语言的
discrete *adj.* 分离的，不相关的	platform *n.* 平台
distinct *adj.* 清晰的，明显的	polymorphism *n.* 多态性
domain *n.* 范围，领域	prediction *n.* 预言
dome *n.* 圆屋顶	premise *n.* 前提
dominant *adj.* 占优势的，突出的	procedure *n.* 程序，过程
efficient *adj.* 有能力的，效率高的	property *n.* 特性，属性
emphasis *n.* 强调，重点	psychology *n.* 心理，心理学
encapsulation *n.* 封装	puppy *n.* 小狗
evolution *n.* 演变，进化，发展	pyrrhic *n.* 出征舞
gigantic *adj.* 巨大的，庞大的	refinement *n.* 精化

robust *adj.* 强壮的，健全的
script *n.* 脚本
segment *n.* 部分，片段
separation *n.* 分离，分开
sequence *n.* 顺序
specification *n.* 说明，详述
specify *vt.* 详述
syntax *n.* 句法，句法规则

temporary *adj.* 临时的，暂时的
thereby *adv.* 由此，因而
thoughtlessly *adv.* 轻率地，草率地
topology *n.* 拓扑结构
trace *vt.* 追踪，发现，找到
tremble *vi.* 发抖，颤动
unambiguous *adj.* 不含糊的，清楚的

Exercises

1. What is planning the performance of a task or an event?
2. What is a step-by-step procedure for solving a problem in a finite amount of time?
3. What is OOP? What is the name of the data structure we used? What are the components of this data structure?
4. What are the characteristics of OOP?
5. Explain the difference between the structured programming and object-oriented programming.
6. Why do we need object-oriented programming?

Chapter 2 Basic Facilities

— *Shifting from C to C++ Programs*

Let's all move one place on.
—Lewis Carroll

Objectives

- To be able to construct a simple C++ program
- To be able to create and recognize legal C++ identifiers
- To understand how to construct programs modularly from small pieces called functions
- To understand the mechanism used to pass information between functions
- To be able to use references to pass arguments to functions

2.1 C++ Program Structure

The C++ program consists of identifiers, declarations, variables, constants, expressions, statements, and comments. A common structure for a simple, one file, C++ program includes the following:

- Comments. They describe program feature, author, and so on.
- Include statements. They specify the header files for libraries.
- Using namespace statement. Typically, this is only "*using namespace std;*".
- Global declarations (constants, types, variables, ...). Avoid global variables if possible.
- Function prototypes (declarations).
- Main function definition.
- Function definitions.

The following example is used to illustrate a C++ program structure. Do not be too concerned with the details in the program—just observe its over-all look and structure.

Example 2-1: A C++ program.

```
//---------------------------------------------------------------------------
//File: example2_1.cpp
//The program reads a given course information and students' grade from the console, calculates
//the average grade and outputs students' grades.
//Liu 2018-07-10
//---------------------------------------------------------------------------
```

```cpp
1    //== includes ==
2    #include <iostream>
3    #include <string>
4    #include <iomanip>
5    using namespace std;
6
7    //declaration of constants
8    const int studentNum = 5;
9
10   //== function prototype ==
11   void setCourseInfo(string &cName, string &lectName, int &cHour);
12   void setGrade(int(&sNo)[studentNum], string (&sName)[studentNum], int(&g)[studentNum]);
13   double average();
14   void printCourseInfor(string, string, int);
15   void printGradeBook(int*, string*, int*);
16
17   //== main function ==
18   int main()
19   {
20      string courseName;
21      string lecturer;
22      int courseHour;
23      int studentNo[studentNum];
24      string studentName[studentNum];
25      int grade[studentNum];
26
27      //Read the course and student data
28      setCourseInfo(courseName, lecturer, courseHour);
29      setGrade(studentNo, studentName, grade);
30
31      //Display the course and student data
32      printCourseInfor(courseName, lecturer, courseHour);
33      printGradeBook(studentNo, studentName, grade);
34      return 0;
35   }
36   //== functions ==
37   //Read the course and student data
38   void setCourseInfo(string &cName, string &lectName, int &cHour)
39   {
40      cout << "Enter course information\n";
41      cin >> cName >> lectName >> cHour;
42   }
43   void setGrade(int(&sNo)[studentNum], string(&sName)[studentNum], int(&g)[studentNum])
44   {
45      cout << "Enter student's grades\n";
46      for (int i = 0; i < studentNum; i++)
47      {
```

```
48        sNo[i] = i;
49        cin >> sName[i] >> g[i];
50    }
51 }
52 void printCourseInfor(string cName, string lectName, int cHour)
53 {
54    cout << "Course name: " << cName << "Course hour: " << cHour
55         << "Lecturer: " << lectName << endl;
56 }
57 void printGradeBook(int* sNo, string* sName, int* sg)
58 {
59    cout << "---------------------------\n";
60    cout << "StudentNo    Name    Grade\n";
61    for (int i = 0; i < studentNum; i++)
62       cout << setw(9) << right << sNo[i] << setw(10) << sName[i]
63            << setw(7) << right << sg[i] << endl;
64    cout << "---------------------------\n";
65    cout << "The average grade: " << average(sg) << endl;
66 }
```

Result:

```
1    Enter course information
2    OOP Liu 48
3    Enter student's grades
4    Wang 95
5    Zhao 78
6    Zhang 60
7    Li 83
8    Qian 45
9    Course name: OOPCourse hour: 48Lecturer: Liu
10   ---------------------------
11   StudentNo      Name      Grade
12       0         Wang        95
13       1         Zhao        78
14       2         Zhang       60
15       3         Li          83
16       4         Qian        45
17   ---------------------------
18   The average grade: 72.2
```

Comments are used for better understanding of the program statements. The comment entries start with two slashes (//) on a line and terminate at the end of the line, *called a **line** comment.* In Example 2-1, the statements

```
//File:example2_1.cpp
//The program reads a given course information and students' grade from the console,
//calculates the average grade and outputs students' grades.
```

Like the C language, a comment entry in a C++ program can be enclosed between /* and */ on one or several lines, called *a **paragraph comment***.

The ***#include*** directive instructs the compiler to include the contents of the file enclosed within angular brackets (< >) into the source file. In Example 2-1, the *iostream* file is included in the example2_1.cpp file.

All C++ programs comprise one or more functions, which are a logical grouping of one or more statements. A ***function*** is identified by a function name and a function body.

A function name is identified by a word followed by parentheses. In Example 2-1, *main* is a function name. All programs must have a function called ***main***. The execution of a program begins with the *main* function. The keyword *int* along with the function name signifies that the function returns an integer value.

A function body is surrounded by curly braces ({}). The braces delimit a block of program statements. Every function must have a pair of braces.

2.2 Input/Output Streams

In Example 2-1, you find a few ***cin*** and ***cout*** statements within the *setCourseInfo* and *setGrade* functions. These statements can get values from the keyboard and output them on the console.

A program performs three basic operations: it gets data, it manipulates the data, and it output the results. Because writing programs for I/O is quite complex, C++ offers extensive support for I/O operations. In C++, I/O is a sequence of characters, called a *stream*, from the source to the destination.

> A **stream** is a sequence of characters from the source to the destination.
> **Input stream** is a sequence of characters from an input device (e.g. keyboard) to the computer.
> **Output stream** is a sequence of characters from the computer to an output device (e.g. monitor).
> 流是从源到目的地的字符序列。
> 输入流是从输入设备（如键盘）到计算机的字符序列。
> 输出流是从计算机到输出设备（如监视器）的字符序列。

2.2.1 Input Stream

The standard library offers ***istream*** for input. The ***istream*** deals with the character string representations of built-in types and can easily be extended to cope with user-defined types.

With input streams, the extraction operator (>>) is used to remove values from the stream.

This makes sense: when the user presses a key on the keyboard, the key code is placed in an input stream. Your program then extracts the value from the stream so it can be used.

The >> operator ("get from") is used as an input operator; *cin* (read see-in) is one of the standard Input/Output streams, i.e. *iostream*. The type of the right-hand operand of >> determines what input is accepted and what is the target of the input operation.

The following statements demonstrate the input stream using *cin*.

```
1   void f( )
2   {
3       int i;
4       cin >> i;       //read an integer into variable i
5       double d;
6       cin >> d;       //read a double-precision, floating-point number into variable d
7   }
```

The statements in Lines 4 and 6 read an integer and a floating-point number, such as 1234 and 34.6, from the standard input into the integer variable *i* and the double-precision, floating-point variable *d*. You may find out that *cin* can read the data directly into any variable type. Using *cin* makes the programmer manipulates data easily and flexibly without worrying about the variable type.

2.2.2 Output Stream

With output streams, the insertion operator (<<) is used to put values in the stream. This also makes sense: you insert your values into the stream, and the data consumer (e.g. monitor) uses them.

The << operator ("put in") are used as an output operation. *cout* (read see-out) is also one of the standard Input/Output streams, i.e. *iostream*. By default, values output to *cout* are converted to a sequence of characters.

The following statements are examples of using *cout*:

```
1   void f( )
2   {
3       double d =5.6;
4       cout << 10 << endl;
5       cout << d << endl;
6       cout << "Hello World!" << endl;
7   }
```

Result:

```
1   10
2   5.6
3   Hello World!
```

Think It Over

What are the differences between the I/O statements of cin/cout and scanf/printf?

2.3 Constants

C++ offers the concept of a user-defined constant, a *const*, to express the notion that the value of a variable cannot be changed directly. This is useful in several contexts. For example, many variables do not actually have their values changed after initialization; symbolic constants lead to more maintainable code than do literals embedded directly in code; pointers are often read through but never written through, and most function parameters are read but not written to. In Examples 1-1 and 1-2, MAX and PI are constant.

The *const* keyword specifies that a variable's value is constant and tells the compiler to prevent the programmer from modifying it. It can be added to the declaration of a variable to declare the variable as a constant.

The syntax for declaring a constant is

Syntax   const dataType constantName = value;

For example,

```
1    const int MaxN = 90;              //Max N is a const
2    const int v[] = {1, 2, 3, 4};     //v[i] is a const
3    const int x;                      //error: no initialization
4    void f()
5    {
6        MaxN = 200;                   //error
7        v[2]++;                       //error
8    }
```

Since a constant cannot be assigned a value, it must be initialized. Declaring a variable as *const* ensures that its value will not change within its scope.

因为一个常量不能赋值,所以它必须被初始化。声明一个标识符为常量,这将确保它的值在其作用范围内不能被修改。

Differences Between the *define* Directive and the *const* Statement

In C++, you can use the ***const*** keyword instead of the ***define*** pre-processor directive to define constant values. Values defined with *const* are subject to type checking and can be used in place of constant expressions. In C++, you can specify the size of an array with a const variable as follows:

```
                const int MaxN = 100;
                char strV[MaxN];                        //allowed in C++; not allowed in C
```

In C, constant values default to the external linkage, so they can appear only in source files. In C++, constant values default to the internal linkage, which allows them to appear in header files.

Depending on how smart it is, a compiler can take advantage of an object being a constant in several ways. For example, the initializer for a constant is often (but not always) a constant expression; if it is, it can be evaluated at compile time. Further, if the compiler knows every use of the const, it does not need to allocate space to hold it.

For example,

```
                const int c1 = 1;
                const int c2 = 2;
                const int c3 = my_f(3);      //don't know the value of c3 at compile time
                extern const int c4;         //don't know the value of c4 at compile time
                const int *p = &c2;          //need to allocate space for c2
```

Given this, the compiler knows the values of $c1$ and $c2$ so that they can be used in constant expressions. Since the values of $c3$ and $c4$ are not known at compile time, storage must be allocated for $c3$ and $c4$. Because the address of $c2$ is taken (and presumably used somewhere), storage must be allocated for $c2$. The simple and common case is the one in which the value of the constant is known at compile time and no storage needs to be allocated; $c1$ is an example of that. The keyword extern indicates that $c4$ is defined elsewhere.

Pointers and Constants

When using a pointer, two variables are involved: the pointer itself and the variable pointed to. "Prefixing" the declaration of a pointer with const makes the variable, but not the pointer, a constant. To declare a pointer itself, rather than the object pointed to, to be a constant, we use the declaratory operator *const* instead of plain*.

Consider the following statements:

```
1       void f(char* p)
2       {
3           char s[] = "Gorm";
4           const char* pc = s;           //pointer to constant
5           pc[3] = 'g';                  //error: pc points to constant
6           pc = p;                       //ok
7
8           char* const cp = s;           //constant pointer
9           cp[3] = 'a';                  //ok
10          cp = p;                       //error
11
```

```
12      const char* const cpc = s;      //const pointer to const
13      cpc[3] = 'a';                   //error
14      cpc = p;                        //error
15  }
```

Note that the statements in Lines 4, 8 and 12 produce different results. Variable *pc* in Line 4 is a pointer to a constant. This means you cannot change the value of variable *pc* whose address it is holding. *cp* in Line 8 is a constant pointer. This means that you cannot change the address pointer *cp* is pointing to, but you can change its value. *cpc* in Line 12 is a constant pointer to a constant. This means that you cannot change neither the address *cpc* is pointing to nor the value whose address *cpc* is holding.

2.4 Functions

Let us look back the program in Example 1-3 of Chapter 1. When you work on a program, it can be divided into a few blocks called ***function body***. Sometimes, it is not practical to put the entire program into one function, e.g. the *main* function, as you see in Example 1-3. You must learn to break the problem into manageable pieces.

Functions are like building blocks. This allows you to divide complicated problems into manageable pieces. The aims for using functions in a program are to:
- be more brief.
- be maintained easily.
- enhance the efficiency of programming.
- be reusable.

2.4.1 Function Declarations

A function ***declaration*** establishes the name of the function, the type of data returned by the function and the number and types of its parameters.

The syntax of a function declaration is

Syntax   returnValueType functionName(parameter list);

A function declaration consists of a return type, a name, and a parameter list. In addition, a function declaration may optionally specify the function's linkage. In C++, the declaration can also specify an exception specification, a const-qualification, or a volatile-qualification.

A declaration informs the compiler of the format and existence of a function prior to its use. A function can be declared several times in a program, provided that all the declarations are compatible. An implicit declaration of a function is not allowed: every function must be explicitly declared before it can be called.

For example,

```
1    double sqrt(double);           //declaration of the sqrt function with a parameter of double type
2    double sr2 = sqrt(2);          //invoke the sqrt function by passing the argument of double(2)
3    double sq3 = sqrt("three");    //error: sqrt() requires an argument of double
```

Formal Parameters and Actual Parameters

When you call a function, the arguments are evaluated, and each parameter is initialized with the value of the corresponding argument. The semantics of the argument passed are identical to those of the assignment.

> A **function argument (i.e. actual parameter)** is an expression or a variable that you use within the parentheses of a function call. A **function parameter (i.e. formal parameter)** is an object or reference declared within the parenthesis of a function declaration or definition.
> 一个函数参数（实参）是在函数调用时参数表括号中使用的一个表达式或一个变量。一个函数参数（形参）是不需要程序员指定的函数的一个参数。

Some declarations do not name the parameters within the parameter lists; the declarations simply specify the types of parameters and the return values. This is also called a *function prototype*. A function prototype consists of the function return type, the name of the function, and the parameter list. For example, the following statement describes such condition:

```
double sqrt(double);
```

Function prototypes are required for compatibility between C and C++. The non-prototype form of a function that has an empty parameter list means that the function takes an unknown number of parameters in C, whereas, in C++, it means that it takes no parameters.

> A **function declaration** consists of its name, return type and parameter list. There maybe exists parameters within the parameter list or not. It is also called a **function prototype**.
> 一个函数声明包括函数名字、返回类型和参数表。在参数表中，可以有参数也可以无参数，函数声明也称为函数原型。

2.4.2 Function Definitions

Every function in a program must be defined somewhere (once only) before it is invoked. A function definition is a function declaration in which the body of the function is presented.

The syntax of a function definition is

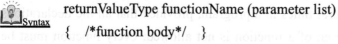

```
returnValueType functionName (parameter list)
{   /*function body*/   }
```

Example 2-2: The declaration and definition of function *swap*.

```
//---------------------------------------------------------------
//File: example2_2.cpp
//This program defines a swap function which exchanges two integers.
//---------------------------------------------------------------
1    #include <iostream>
2    using namespace std;
3
4    void swap (int*, int*);              //function declaration
5
6    void swap (int *p, int *q)           //function definition
7    {
8        int t = *p;
9        *p = *q;
10       *q = t;
11   }
12   int main()
13   {
14       int a = 3, b = 5;
15       cout << "a = " << a << " b = " << b << endl;
16       swap(a, b);
17       cout << "a = " << a << " b = " << b << endl;
18       return 0;
19   }
```

The types of definitions and declarations for a function must specify the same type. The parameter names, however, are not part of the type and not needed to be identical. It is uncommon to have function definitions with unused parameters.

```
void swap (int *p, int *g, char*)
{
    //no use of the third parameter
}
```

As shown above, the fact that a parameter is unused can be indicated by not naming it. Typically, unnamed parameters arise from the simplification of code or from planning ahead for extensions. In both cases, leaving the parameter in place, although unused, ensures that callers are not affected by the change.

A function **definition** consists of its name, its return type, its parameter list and body.

The function **header** specifies the function's return type, name and parameters.

The function **signature** consists of the function name and the parameter list.

一个函数定义由函数名字、返回类型、参数表和函数体构成。

一个函数头包括函数名字、返回类型、参数。

一个函数签名由函数名字和参数表构成。

2.4.3 Default Arguments

A general function often needs more parameters that are necessary to handle simple cases. In particular, functions that construct objects often provide several options for flexibility.

> A **default argument** is an argument to a function that a programmer is not required to specify.
> 默认（缺省）参数是函数的一个参数，程序员没有要求指定的。

A ***default argument*** means that the default values are passed to the parameters when a function is invoked without the arguments. Of course, the user can change the default value if he or she provides a value for the default parameter.

Consider a function for printing an integer in Example 2-3. Giving the user an option of what *value*2 to be printed in seems reasonable, but integers will be printed as decimal integer values in most programs.

Example 2-3: A *print* function using default arguements.

```
//--------------------------------------------------------
//File: example2_3.cpp
//This program defines a print function with default arguments.
//--------------------------------------------------------
1    #include <iostream>
2    using namespace std;
3
4    void print (int value1, int value2 = 10);   //default value of parameter value2 is 10
5
6    int main ()
7    {
8        print(31);
9        print(32, 30);
10       return 0;
11   }
12
13   void print(int value1, int value2)
14   {
15       cout << value1 << "   " << value2 <<endl;
16   }
```

Result:

```
1    31   10
2    32   30
```

In the first function call in Line 8, the caller does not supply an argument for variable *value*2, so the function uses the default value of 10. In the second call in Line 9, the caller does supply a value for variable *value*2, so the user-supplied value is used.

· 28 ·

A default argument is checked for type at the time of the function declaration and evaluated at the time of the call. Default arguments may be provided for trailing parameters only. For example, here is a function prototype for which multiple default arguments might be commonly used.

Example 2-4: A *print* function with multiple default arguments.

```
//----------------------------------------------------------------
//File: example2_4.cpp
//This program defines a print function with multiple default arguments.
//----------------------------------------------------------------
1    #include <iostream>
2    using namespace std;
3
4    void print (int value1 = 10, int value2 = 20, int value3 = 30);
5    int main()
6    {
7        print();
8        print(30);
9        print(30, 40);
10       print(30, 40, 50)
11       return 0;
12   }
13   void print(int value1, int value2, int value3)
14   {
15       cout << value1 << " " << value2 << " " << value3 << endl;
16   }
```

Result:

```
1   10   20   30
2   30   20   30
3   30   40   30
4   30   40   50
```

In Example 2-4, you might notice the statements in Lines 4 and 13. They indicate the function declaration with default values in Line 4 and function definition without default values in Line 13.

It is impossible to supply a user-specified value for *value3* without also supplying a value for *value1* and *value2*. The reason is that C++ does not support a function call like *print*(, , 3).

When you use the default argument, there are two major consequences to be obeyed.
- All default parameters within declaration or definition must be the rightmost parameters.

The following is not allowed:

void print(int value1 = 0, int value2= 0, int value3);   //error

void print(int value1 = 0, int value2, int value3 = 0);	//error
void print(int value1, int value2 = 0, int value3 = 0);	//ok

- The leftmost default parameter should be the one most likely to be changed by the user when the function is called.

当你使用默认参数时，要遵循两个要点：

- 在函数声明或定义中所有的默认参数必须遵循最右原则，但不允许下列情况：

void print(int value1 = 0, int value2= 0, int value3);	//error
void print(int value1 = 0, int value2, int value3 = 0);	//error
void print(int value1, int value2 = 0, int value3 = 0);	//ok

- 当函数调用时，最左默认参数的值应该最先被用户改变。

2.4.4　Inline Functions

An ***inline function*** is one for which the compiler copies the code from the function definition directly into the code of the calling function rather than creating a separate set of instructions in memory. Instead of transferring control to and from the function code segment, a modified copy of the function body may substitute the function directly for the function call. In this way, the performance overhead of a function call is avoided.

A function is declared inline by using the inline function specifier or by defining a member function within a class or structure definition. The inline specifier is only a suggestion to the compiler that an inline expansion can be performed; the compiler is free to ignore the suggestion.

The following statements show the inline function definition.

```
inline int maxValue (int n, int m)
{
    return (n < m) ? m : n;
}
```

The inline specifier is a hint to the compiler that it should attempt to generate code for a call of function *maxValue* inline rather than laying down the code for the function once and then calling through the usual function call mechanism.

The use of the inline specifier does not change the meaning of the function. However, the inline expansion of a function may not preserve the order of evaluation of the actual arguments. Inline expansion neither changes the linkage of a function: the linkage is external by default.

2.4.5　Function Overloading

Most often, it is a good idea to give different functions different names, but when some functions conceptually perform the same task on objects of different types, it can be more convenient to give them the same name. Using the same name for operations on different types is called ***overloading***.

The technique is already used for the basic operations in C++. That is, there is only one name for addition, +, yet it can be used to add values of integer, floating-point, and pointer types. This idea is easily extended to functions defined by the programmer. Several functions have the same name. This is called overloading a function name or ***function overloading***.

Example 2-5: Overloading functions.

```
//-----------------------------------------------------------------
//File: example2_5.cpp
//This program defines three print functions with the same name, each performing a similar
//operation on a different data type.
//-----------------------------------------------------------------
1   #include <iostream>
2   using namespace std;
3
4   void print(int i)
5   {
6       cout << " Here is int " << i << endl;
7   }
8   void print(double f)
9   {
10      cout << " Here is double " << f << endl;
11  }
12
13  void print(char* c)
14  {
15      cout << " Here is char* " << c << endl;
16  }
17
18  int main()
19  {
20      print(10);
21      print(10.10);
22      print("ten");
23      return 0;
24  }
```

Result:

```
1   Here is int 10
2   Here is double 10.10
3   Here is char* ten
```

These *print* functions in Example 2-5 have different parameter lists. Function *print* in

Line 4 takes one parameter of type *int*. Function *print* in Line 8 takes one parameter of type *double*. Function *print* in Line 13 takes one parameter of type char*. The parameter types of these functions are different. They all overload the function name *print*.

 To overload a function name, any two definitions of functions must have different formal in parameter lists:
• Different numbers of formal parameters, or
• If the number of formal parameters is the same, data type of the formal parameters, in the order you list them, must differ in at least one position.

要重载一个函数名，任何两个函数的定义在参数列表中必须有不同的形参：
• 形参的个数不同，或
• 如果形参的数目相同，那么形参的数据类型在你列出它们的顺序中，必须至少在一个位置上不同。

How to Match Function Calls with Overloaded Functions

Given that the following overloading functions are defined.

```
void print (int);
void print(const char*);
void print(double);
void print(long);
void print(char);
```

An example is given by using the above definitions.

```
//--------------------------------------------------------------------------------
//This program calls overloaded functions with different variable types.
//--------------------------------------------------------------------------------
1   void main()
2   {
3       char c='a';
4       int i = 5;
5       short s =7;
6       float f =5.67;
7
8       print(c);            //exact match: invoke print(char)
9       print(i);            //exact match: invoke print(int)
10      print(s);            //short to int promotion: invoke print(int)
11      print(f);            //float to double promotion: print(double)
12      print('a');          //exact match: invoke print(char)
13      print(49);           //exact match: invoke print(int)
14      print(0);            //exact match: invoke print(int)
15      print("a");          //exact match: invoke print(const char*)
16  }
```

The call *print*(0) invokes *print(int)* because 0 is an *int*. The call *print('a')* invokes *print (char)* because 'a' is a *char*. The reason to distinguish between conversions and promotions is that we prefer safe promotions, such as *char* to *int*, over unsafe conversions, such as *int* to *char*.

Therefore, making a call to an overloaded function results in one of the three possible outcomes:

(1) A match is found. The call is resolved to a particular overloaded function.

(2) No match is found. The arguments cannot be matched to any overloaded function.

(3) An ambiguous match is found. The arguments matched more than one overloaded function.

When an overloaded function is called, C++ goes through the following process to determine which version of the function will be called.

(1) C++ tries to find an exact match. This is the case where the actual parameter matches the parameter type of one of the overloaded functions exactly. For example, in the example above,

```
print(c);
print(i);
print(s);
print(0);
```

Although integer 0 could technically match *print(char*)*, it exactly matches *print(int)*. Thus *print(int)* is the best match available.

(2) If no exact match is found, C++ tries to find a match through promotion, that is, certain types can be automatically promoted via internal type conversion to other types. The conversions among different types are:

• *Char*, *unsigned char,* and *short* is promoted to an *int*.

• *Unsigned short* can be promoted to *int* or unsigned *int*, depending on the size of an *int*.

• *Float* is promoted to *double*.

• *Enum* is promoted to *int*.

For example,

```
1    void print(char* );
2    void print(int);
3
4    print('a')    //promoted to match print(int)
```

In this case, as there is no *print(char)*, the char 'a' is promoted to an integer, which then matches *print(int)*.

(3) If no promotion is found, C++ tries to find a match through standard conversion. Standard conversions include:

- Any numeric type will match any other numeric type, including unsigned (e.g. *int* to *float*).
- Enum will match the formal type of a numeric type (e.g. *enum* to *float*).
- Zero will match a pointer type and numeric type (e.g. 0 to *char**, or 0 to *float*).
- A pointer will match a void pointer.

For example,

```
1    void print(struct );
2    void print(float);
3
4    print('a')      //promoted to match print(float)
```

In this case, because there is no *print(char)*, and no *print(int)*, the 'a' is converted to a *float* and matched with *print(float)*.

Ambiguous Matches

If every overloaded function has to have unique parameters, how is it possible that a call could result in more than one match? Because all standard conversions are considered equal, and all user-defined conversions are considered equal, if a function call matches multiple candidates via standard conversion or user-defined conversion, the result will be an ambiguous match.

For example,

```
1    void print(unsigned int );
2    void print(float);
3
4    print('a')      //error: ambiguous match
```

In the case of *print*('a'), C++ cannot find an exact match. It tries promoting 'a' to an *int*, but there is no *print*(int) neither. Using a standard conversion, it can convert 'a' to both an *unsigned int* and a *floating-point* value. Because all standard conversions are considered equal, this is an ambiguous match.

Default Arguments and Overloading Functions

Sometimes, functions with default parameters may be overloaded.

For example, the following is allowed:

```
1    void print(char *sPtr );
2    void print(char c = ' ');
3    print()         //ok: default parameter
4    print(' ');     //ok: match print(char c=' ')
```

In this case, the statement in Line 3 is executed by calling the *print* function in Line 2. The statement in Line 4 is executed by calling the *print* function in Line 1.

However, it is important to note that default parameters do NOT count towards the parameters that make the function unique. Consequently, the following is not allowed:

```
1    void print(int value );
2    void print(int value1, int value2 = 10);
3
4    print(10);            //error: ambiguous
```

If the caller were to call *print*(10), the compiler would not be able to disambiguate whether the user wanted *print(int)* or *print(int, 20)* with the default value.

Think These Over

1. What is the difference between a function declaration and a function definition?

2. What is the purpose of using default values in the formal parameter list of the function?

3. If the functions have the same name and formal parameter list, but they have different return types, do you think the function name can be overloaded?

2.5 References

2.5.1 Reference Definition

> A **reference** is a simple reference data type. A reference is an alias or alternative name for an object. This means that C++ allows you to create a second name for the object that you can use to read or modify the original data stored in the object.
> 引用是一种简单的引用数据类型。引用是一个对象的别名或另一个名字。这就意味着C++允许你为这个对象创建第二个名字，可用第二个名字来读取或修改这个对象存储的原始数据。

The main use of references is for specifying arguments and return values for functions in general and for overloaded operators in particular.

The syntax of declaring a reference variable is

Syntax   variableType& variableName;

where *variableType* is a data type and *variableName* is an identifier whose type is reference to *variableType*.

For example,

int i = 1;	
int &r = i;	//variables *r* and *i* now refer to the same *int*
int x = r;	//*x* = 1
r = 2;	//*i* = 2
int &r1;	//error

Notice that the statement *int&r = i* means that variable *r* is a reference type. When variable *r* is declared, it will become an alias for integer variable *i*, that is, the address of variable *r* is the same as that of variable *i*. Whenever you use variable *r*, you can just treat it as though it were a regular integer variable. But when you create it, you must initialize it with variable *i*.

Example 2-6: Definition of reference variable *r*.

```
//---------------------------------------------------------------
//File: example2_6.cpp
//This program defines the reference variable r within the main function.
//---------------------------------------------------------------
1   #include <iostream>
2   using namespace std;
3
4   int main()
5   {
6       int iOne;
7       int &r = iOne;
8       iOne = 5;
9
10      cout << "iOne: " << iOne << endl;
11      cout << "r: " << r <<endl;
12      cout << "&iOne: " << &iOne << endl;
13      cout << "&r: " << &r << endl;
14
15      int iTwo = 8;
16      r = iTwo;
17
18      cout << "iOne: " << iOne << endl;
19      cout << "iTwo: " << iTwo << endl;
20      cout << "r: " << r << endl;
21
22      cout << "&iOne: " << &iOne << endl;
23      cout << "&iTwo: " << &iTwo << endl;
24      cout << "&r: " << &r << endl;
25      return 0;
26  }
```

Result:

```
1   iOne: 5
2   r: 5
3   &iOne: 0065FDF4
4   &r: 0065FDF4
5   iOne: 8
6   iTwo: 8
7   r: 8
8   &iOne: 0065FDF4
9   &iTwo: 0065FDEC
10  &r: 0065FDF4
```

In this example, let us assume that memory location 0065FDF4 is allocated for *iOne*, and memory location 0065FDEC is allocated for *iTwo*, as shown in Figure 2-1.

Here, *r* is of type "reference to *int*". The *r* variable refers to an integer variable *iOne*. This means that both variables *r* and *iOne* have the same address in the memory. The *r* variable is changed as *iOne* is changed. The statement in Line 16 assigns *r* by variable *iTwo*. We find that the value of *iOne* is changed (see Line 5 of Result).

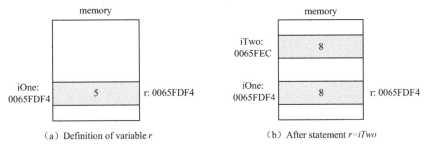

(a) Definition of variable *r*　　　　　　(b) After statement *r=iTwo*

Figure 2-1　Variable *r* in Example 2-6

References and Pointers

We can find out that a reference variable is used easily as a regular variable. Notice that the reference variables do not need to be allocated a new address in the memory. From this point, we have to consider a pointer variable. The following example illustrates the difference between references and pointers.

Example 2-7: Definitions of reference *r* and pointer *ipt*.

```
//---------------------------------------------------------------
//File: example2_7.cpp
//This program demonstrates the difference between reference r and pointer ipt.
//---------------------------------------------------------------
1   #include <iostream>
2   using namespace std;
3
4   int main()
5   {
6       int i1;
7       int& r = i1;            //a reference
8       int* ipt = &i1;         //a pointer
9       i1 = 5;
10      cout << "r=" << r << endl;
11      cout << "&i1=" << &i1 << endl;
12      cout << "&r=" << &r << endl;
13      cout << "*ipt=" << *ipt << endl;
14      cout << "ipt=" <<ipt<< endl;
15      cout << "&ipt=" <<&ipt<< endl;
16      cout<<"---------------------\n";
17      int i2 = 8;
```

```
18    r = i2;
19    ipt = &i2;
20    cout << "r=" << r << endl;
21    cout << "&i1=" << &i1 << endl;
22    cout << "&i2=" << &i2 << endl;
23    cout << "&r=" << &r << endl;
24    cout << "*ipt=" << *ipt << endl;
25    cout << "ipt=" << ipt << endl;
26    cout << "&ipt=" << &ipt << endl;
27
28    return 0;
29  }
```

Result:

```
1    r=5
2    &i1=0012FF7C
3    &r=0012FF7C
4    *ipt=5
5    ipt=0012FF7C
6    &ipt=0012FF74
7    -----------------------
8    r=8
9    &i1=0012FF7C
10   &i2=0012FF70
11   &r=0012FF7C
12   *ipt=8
13   ipt=0012FF70
14   &ipt=0012FF74
```

In this example, *r* is of type "reference to *int*", and *ipt* is of type "pointer to *int*". The value of reference *r* is an integer, but the value of pointer *ipt* is a memory address, Figure 2-2 shows their values in the memory.

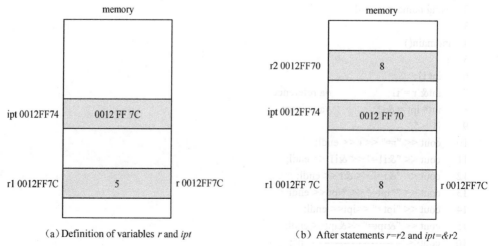

Figure 2-2   Variables *r* and *ipt* in Example 2-7

2.5.2 Reference Variables as Parameters

Parameters provide a communication link between the calling function (such as *main*) and the called function. They enable functions to manipulate different data each time they are called. In general, there are two types of formal parameters: value parameters and reference parameters.

> A **value parameter** is a formal parameter that receives a copy of the content of the corresponding actual parameter.
>
> A **reference parameter** is a formal parameter that receives the location (memory address) of the corresponding actual parameter.
>
> 当**值参数**是形参时,它将收到对应实参内容的一个副本。
>
> 当**引用参数**是形参时,它将收到对应实参的位置(内存地址)。

The following program shows how a value parameter and a reference parameter work.

Example 2-8: Reference variables as parameters.

```cpp
//-------------------------------------------------------------------
//File: example2_8.cpp
//This program defines two functions, having a reference parameter and a value parameter.
//-------------------------------------------------------------------
1   #include <iostream>
2   using namespace std;
3
4   int byValue(int);              //passing arguments by value
5   int byReference(int&);         //passing arguments by reference
6
7   int main()
8   {
9       int x = 2;
10      int y = 4;
11      cout << byValue( x ) << endl;
12      cout << "x = " << x << endl;
13      cout<< byReference( y );
14      cout << "y = " << y << endl;
15      return 0;
16  }
17  int byValue(int number)
18  {
19      return number *= number;
20  }
21  int byReference(int& number)
22  {
23      return number *= number;
24  }
```

Result:

```
1    4
2    x = 2
3    16
4    y = 16
```

The statement in Line 11 calls the function *byValue*. The value of the actual parameter *x* is then passed to a formal parameter *number*. After the statement executes, the function *byValue* returns the value of *number* *= *number*. When the function executes, any changes made to the formal parameter *number* do not in any way affect the actual parameter *x*. The actual parameter *x* has no knowledge of what is happening to the formal parameter *number*. So the statement in Line 12 outputs 2 for *x*. Thus, value parameters provide a one-way link between actual parameters and formal parameters. Hence, functions with value parameters have limitations.

The statement in Line 13 calls the function *byReference*. The value of the actual parameter is passed to a formal parameter *number*. Likewise, the function *byReference* returns the value of *number* *= *number* after the statement executes. Because the formal parameter *number* is a reference parameter, it receives the address (memory location) of the actual parameter *y*. Thus, reference parameter *number* can pass one value for the function and can change the value of the actual parameter *y*. So the statement in Line 14 outputs 16 for *y*.

Reference parameters are useful in three situations:
- when you want to return more than one value from a function;
- when the value of the actual parameter needs to be changed;
- when passing the address would save memory space and time relative to copying a large amount of data.

引用参数在下列三种情形下非常有用：
- 当你想从函数中返回多值时；
- 当函数实参的值需要改变时；
- 当复制大量数据时，传地址可以节省空间和时间。

2.5.3　References as Returning Values

Returning values from a function to its caller by value, address, or reference works almost the same way as passing parameters to a function. They have the same advantages and disadvantages. The primary difference between the two is simply the opposite directions of data flow. However, there is one more added bit of complexity—because local variables in a function go out of scope when a function returns, we need to consider the effect of this on each return type.

Just like pass by reference, values returned by reference must be variables. When a variable is returned by reference, a reference to the variable is passed back to the caller. The caller can then use this reference to continue modifying the variable, which can be useful at times. Return by reference is also fast, which becomes useful when returning structs and classes.

Example 2-9: References as returning values.

```
//----------------------------------------------------------------
//File: example2_9.cpp
//This program defines two functions which return by reference and value.
// ----------------------------------------------------------------
1   #include <iostream>
2   using namespace std;
3
4   int returnValue(int n)
5   {
6       int nValue = n * 2;
7       return nValue;        //a copy of nValue will be returned here
8   }  //nValue goes out of scope here
9
10  int& returnReference(int n)
11  {
12      int nValue = n * 2;
13      return nValue;        //return a reference to nValue here
14  }  //nValue goes out of scope here
15
16  int main()
17  {
18      int x = 4;
19      cout << " returnValue = " << returnValue(x) << endl;
20      cout << "returnReference = " << returnReference(x) <<endl;
21      return 0;
22  }
```

Result:

```
1   returnValue = 8
2   returnReference = 8
```

Returning a value is the simplest and safest return type to use. When a value is returned by value, a copy of that value is returned to the caller.

Passing (or returning) a value or a reference is considered in the following cases:
- As with passing by value, you can return by value literals (e.g. 5), variables (e.g. *x*), or expressions (e.g. *x* + 1), which makes a return by value very flexible.
- As with passing by reference, you cannot return a reference to a literal or expression.

- As with returning by value, you can return a variable (or expression) that involves local variable *nValue* declared within function *returnValue*. Because the variable is evaluated before the function goes out of scope, and a copy of *nValue* is returned to the caller, there is not a problem when the variable goes out of scope at the end of the function.
- As with returning by reference, the problem within the function *returnReference*—is trying to return value *nValue*, which goes out of scope when the function returns, passed by reference to the function and back to the caller. To solve this problem, we can declare *nValue* to be a global variable.

在下列情形下，考虑使用传递（或返回）值或引用参数：
- 对于传递值参数，可以通过值文字（例如 5）、变量（例如 x）或表达式（例如 x+1）返回，这使得按值返回非常灵活。
- 对于传递引用参数，不能返回对文字或表达式的引用。
- 对于按值返回，则是返回一个变量（或表达式），它是在函数 returnValue 中声明的局部变量 nValue。因为变量是在函数作用域范围内求值的，并且将 nValue 的副本返回给调用者，所以当变量脱离函数作用域时没有问题。
- 对于按引用返回，函数 returnReference 中则出现问题，当函数返回时，通过传递引用试图返回 nValue 值给函数调用者，但该变量值脱离了函数作用域范围。为了解决这个问题，我们可以声明 nValue 是全局变量。

2.5.4　Reference as Left-Hand Values

Returning by reference is typically used to return arguments passed by reference to the function back to the caller—used on the right-hand side of an assignment operator. If the function call can appear on the left-hand side of an assignment operator, what will happen? In the following example, we will return (by reference) an element of an array that was passed to our function by reference on the left-hand of an assignment operator.

> A **left-hand value** is a value on the left-hand side of the assignment operator.
> 左值是在赋值运算符左侧的值。

Example 2-10: Reference as a left-hand value.

```
//--------------------------------------------------------------------------------
//File: example2_10.cpp
//This program assigns a value to the element of an array using left-hand value.
```

```
//------------------------------------------------------------
1   #include <iostream>
2   using namespace std;
3
4   int a[10];
5
6   int & Arr(int i);
7   {   return a[i]; }
8
9   int main()
10  {
11      Arr(4) = 50;      //left-hand
12      cout << a[4];
13      return 0;
14  }
```

Result:

```
1   50
```

When we call *Arr(4)* in Line11, function *Arr* returns a reference to the 4th element of the *a* array. The *main* function then uses this reference to assign the 4th element the value 50.

Although this is somewhat of a contrived example (because you could access a directly), once you learn about classes, you will find a lot more uses for returning values by reference.

Think These Over

1. What are the differences between pointers and references?
2. What is the principal reason for passing parameters by reference?

2.6 Namespaces

A ***namespace*** is a mechanism for expressing logical grouping, i.e. if declarations logically belong together according to specific criteria, then a common namespace can be used to express this fact.

The syntax of declaring namespaces is

```
namespace identifier
{
    entities
}
```

where *identifier* is any valid identifier and entities is the set of classes, objects and functions that are included within the namespace. For example,

```
namespace mySpace {
    int a, b;
```

```
        double f, d;
}
namespace yourSpace {
    int a, b;
    double f, d;
}
```

In this example, variables *a*, *b*, *f* and *d* are declared within two namespaces called *mySpace* and *yourSpace* respectively. Although these variables have the same name, declaring them within the two namespaces will not cause a redefinition error; the namespace is able to avoid **name collisions** (of variables, types, classes or functions).

In order to access these variable from the *mySpace* and *yourSpace* namespaces, we have to use the scope operator (::). For example, to access variable a from outside the namespaces, we can write,

```
mySpace::a
yourSpace::a
```

Example 2-11: The definition of two namespaces.

```
//--------------------------------------------------------------------------
//File: example2_11.cpp
//This program defines two namespaces, mySpace and yourSpace.
//--------------------------------------------------------------------------
1   #include <iostream>
2   using namespace std;
3
4   namespace mySpace
5   {
6       int var = 5;
7   }
8
9   namespace yourSpace
10  {
11      double var = 3.1416;
12  }
13
14  int main()
15  {
16      cout << mySpace::var << endl;
17      cout << yourSpace::var << endl;
18      return 0;
19  }
```

Result:

```
1   5
2   3.1416
```

The ***using*** keyword is used to introduce a name from a namespace into the current declarative region.

Example 2-12: Utilizing the ***using*** keyword for namespaces.

```
//----------------------------------------------------------------------------
//File: example2_12.cpp
//This program utilizes the using keyword for namespaces.
//----------------------------------------------------------------------------
1   #include <iostream>
2   using namespace std;
3   namespace mySpace
4   {
5       int a = 5;
6       int b = 6;
7   }
8   namespace yourSpace
9   {
10      double a = 3.1416;
11      double b = 5.678;
12  }
13  int main()
14  {
15      using mySpace::a;
16      using yourSpace::b;
17      cout << a << endl;
18      cout << b << endl;
19      cout << mySpace::b << endl;
20      cout << yourSpace::a << endl;
21      return 0;
22  }
```

Result:

```
1   5
2   5.678
3   6
4   3.1416
```

Notice how in this example, variable *a* (without any name qualifier) refers to *mySpace::a* in Line 17, whereas variable *b* refers to *yourSpace::b* in Line 18, exactly as our using declarations have specified. We still have access to *mySpace::b* and *yourSpace::a* using their fully qualified names.

Example 2-13: The same name functions in namespaces.

```
//----------------------------------------------------------------------------
//File: example2_13.cpp
//This program defines the same name variables and functions within two namespaces.
```

```
//  ------------------------------------------------------------------------
1   #include <iostream>
2   using namespace std;
3
4   namespace mySpace
5   {
6       int a = 5;
7       int b = 6;
8       int add(int, int);
9   }
10
11  namespace yourSpace
12  {
13      double a = 3.1416;
14      double b = 5.678;
15      double add(double, double);
16  }
17
18  int mySpace::add(int x, int y) { return x + y; }
19  double yourSpace::add(double x, double y) { return x + y; }
20
21  int main()
22  {
23      using namespace mySpace;
24      cout << a << endl;
25      cout << b << endl;
26      cout << add(a, b);
27
28      std::cout << yourSpace::a << endl;
29      std::cout << yourSpace::b << endl;
30      std::cout << yourSpace::add (yourSpace::a, yourSpace::b) << endl;
31
32      return 0;
33  }
```

Result:

```
1   5
2   6
3   11
4   3.1416
5   5.678
6   8.8196
```

In this example, we define the function add within the two namespaces respectively. Since we have declared that we were using the statement *using namespace mySpace* in Line 23, all direct uses of the *a* and *b* variables, and the function add without name qualifiers (in Lines 24, 25 and 26) are referring to their declarations in namespace *mySpace*. So, we can use the variables and functions within the namespace *yourSpace*.

> Here the function bodies are defined outside the namespaces *mySpace* and *yourSpace*, the aim of this is to separate the interface (i.e. function declaration) from the implementation (i.e. function definition).
>
> As a result of separating the implementation of the interface, each function now has exactly one declaration and one definition. Users will see only the interface containing declarations. The implementation—in this case, the function bodies—will be places "somewhere else" where a user does not need to see.
>
> 这里将函数体定义在名称空间 mySpace 和 yourSpace 外，其目的是将接口（函数声明）与实现（函数定义）分开。
>
> 由于分开了接口的实现，每个函数现在都有一个声明和一个定义。用户只会看到包含声明的接口。在这种情况下，函数体的实现将被放置在用户不需要看的地方。

You may notice the ***std*** specifier is used in Example 2-13. Because all the files in the C++ standard library declare all of its entities within the std namespace, we have generally included the *using namespace std;* statement in all programs that used any entity defined in iostream.

Word Tips

allocate *vt.* 分配	evaluate *vi.* 赋值，评估
argument *n.* 参数，自变量	execution *n.* 执行
array *n.* 数组	extensive *adj.* 广泛的
actual parameter 实参	flexibility *n.* 灵活性
bracket *n.* 括号	formal parameter 形参
built-in *adj.* 系统建立的，内置的	hint *vt.* 暗示，示意
comment *n.* 注释	identical *adj.* 同一的，相同的
compatibility *n.* 兼容性	identifier *n.* 标识符
comprise *vt.* 包含	illustration *n.* 举例说明
context *n.* 上下文	implementation *n.* 实现
contrived *adj.* 不自然的，做作的	implicit *adj.* 隐含的
corresponding *adj.* 相应的，相关的	interface *n.* 接口，界面
criteria *n.* 标准	linkage *n.* 链接
curly braces *n.* 花括号	manipulate *v.* 计算，处理
decimal *adj.* 小数的，十进制的	mechanism *n.* 机制
elaborate *vi.* 变复杂的	multiple *adj.* 许多的
embedded *adj.* 嵌入的	notion *n.* 概念
enum *n.* 枚举	optionally *adv.* 任意地

parentheses（复数） *n.* 圆括号
presumably *adv.* 大概
prototypes *n.* 原型
semantics *n.* 语义学，语意论
substitute *adj.* 代替的

syntax *n.* 语法，句法
trailing *adj.* 拖尾的，尾部的
variable *n.* 变量
volatile *adj.* 易变的

Exercises

1. Answer the following questions:

(1) Dividing a program into functions

a. is the key to object-oriented programming.

b. makes the program easier to conceptualize.

c. may reduce the size of the program.

d. makes the program run faster.

(2) A function's single most important role is to

a. give a name to a block of code.

b. reduce program size.

c. accept arguments and provide a return value.

d. help organize a program into conceptual units.

(3) A default argument has a value that

a. may be supplied by the calling program.

b. may be supplied by the function.

c. must have a constant value.

d. must have a variable value.

(4) A function argument is

a. a variable in the function that receives a value from the calling program.

b. a way that functions resist accepting the calling program's values.

c. a value sent to the function by the calling program.

d. a value returned by the function to the calling program.

(5) Overloaded functions

a. are a group of functions with the same name.

b. all have the same number and types of arguments.

c. make life simpler for programmers.

d. may fail unexpectedly due to stress.

(6) When an argument is passed by reference

a. a variable is created in the function to hold the argument's value.

b. the function cannot access the argument's value.

c. a temporary variable is created in the calling program to hold the argument's value.

d. the function accesses the argument's original value in the calling program.

2. Read the following program and complete it:

```
#include <iostream>
using namespace std;
int& put(int n);        //put value into the array
int get(int n);         //obtain a value from the array
int vals[10];
int error = -1;
void main()
{
    put(0) = 10;        //put value into the array
    put(1) = 20; put(2) = 30;
    cout<< get(0)<<endl;
    cout<< get(1)<<endl;
    cout<< get(2)<<endl;
    put(12) = 1;        //out of bound
    cout<<get(12)<<endl;
}
```

3. Write out the output of the following programs.

(1)

```
int i = 7;
int& r = i;
r = 9;
i = 10;
cout<<r<<" ; "<<i<<endl;
cout<<&r<<" ; "<<&i<<endl;
```

(2)

```
namespace mySpace{
    int a = 5, b = 6;
}
namespace yourSpace{
    double a = 4.5, b = 98.3;
}
void main()
{
    using mySpace::a;
    using yourSpace::b;
    cout<<a<<";"<<b<<endl;
}
```

(3)
```
#include <iostream>
using namespace std;
void Test(int&, int);
void main()
{
    int x = 12;
    int y = 14;
    Test(x, y);
    cout<<"After the first call of Test, the variables equal "<< x << y << endl;
    x = 20;
    y = 22;
    Test(x, y);
    cout<<"After the second call of Test, the variables equal "<< x << y << endl;
}
void Test(int& a, int b)
{
    a = 3;
    a += 2;
    b = a *5;
    cout << "In function Test, the variable equal " << a << b << endl;
}
```

4. Write a function that swaps two integer numbers and two double floating-point numbers. Hint: using *reference and pointer* as the parameters.

5. Write a program that has three overloading functions *display* to produce their outputs. The first function returns a *double* type. The second function returns an *int* type. The third function returns a *char* type.

6. Write a function that takes an integer value and returns the number with its digits reversed. For example, given the number 1234, the function should return 4321.

7. Write a function *qualityPoints* that inputs a student's average and a returning value—4 if a student's average is 90-100, 3 if the average is 80-89; 2 if the average is 70-79, 1 if the average is 60-69, and 0 if the average is lower than 60.

8. Write a function that finds the maximum and minimum values of an array.

9. Write a function *integerPower (base, exponent)* that returns the value of base^{exponent}. For example, integerPower(3, 4)=3*3*3*3. Assume the exponent is a positive, nonzero integer and that base is an integer. (Do not use any math library function)

10. Write a function that inputs 10 integer numbers and outputs the sum of numbers (>=0) and the sum of numbers (<0).

11. Write a function to calculate the calorie of lunch. The user inputs three lines. Each line has a character (*M* is for meat, *V* is for vegetable, *D* is for dessert) and a calorie value. The

program outputs the total calorie value of lunch.

12. According to the input string, count the number of digit and alphabet and output them.

13. Write a function to solve roots of $ax^2+bx+c=0$ equation. Hint: for different a, b and c, the equation may have 0, 1 or 2 roots. You may define a function as follows:

int *solving*(double a, double b, double c, double& x1, double& x2)

where parameters *x1* and *x2* are the roots. The returning value of the function is the number of roots.

14. Write a program that plays the game "guess the number" as follows. Your program chooses the number to be guessed by selecting an integer randomly in the range 1 to 1000. The program then displays the following:

"I have a number between 1 and 1000. Can you guess my number?

Please type your number: "

The player then types a guess. The program responds with one of the following after every guess:

(1) Bingo! You guessed the number.

(2) Too high. Try again…

(3) Too low. Try again…

And the program counts the number of guesses that the player makes.

(1) If the number is 5 or fewer, print "You know the secret?"

(2) If the player guesses the number in 10 tries, then print "You got lucky!"

(3) If the player makes more than 10 guesses, then print "You should be able to do better."

Chapter 3 Foundation of Classes and Objects

—*Data Abstraction and Definition of Classes*

> *Private faces in public places are wiser and nicer than public faces in private places.*
> —*W.H.Auden*

Objectives
- To understand data abstraction and encapsulation
- To be able to create a user-defined type, namely class
- To understand how classes are implemented
- To be able to define constructors and destructors
- To be able to use objects

3.1 Introduction to Structures

A structure type is a user-defined type. Structures are aggregate data type built by using elements of other types, that is, structures are a way of storing many different values in variables of potentially different types under the same name. This makes it a more modular program, which is easier to modify because its design makes things more compact. For example, a structure can be used to store information about students in a database.

When you define a structure, you must remember that the main difference between structures and arrays is that the structure is a collection of different data types and the array is a collection of the same data type.

3.1.1 Defining a Structure in C++

The structure is defined by using the ***struct*** keyword followed by the structure name. Consider the following structure definition:

```
struct Date{
    int day;       //1-31
    int month;     //1-12
    int year;      //1-9999
};
```

The ***struct*** keyword introduces the structure definition. The *Date* identifier is the structure tag that names the structure definition and is used to declare variables of the structure type. In this example, the new type name is *Date*. The names declared in the braces of structure definition are the structure's members. Members of the same structure must have unique names, but two different structures may contain members of the same name without conflict. Each structure definition must end with a semicolon.

The preceding structure definition does not reserve any space in memory, rather, the definition creates a new data type that is used to declare a variable. Structure's variable is declared like variables of other types. For example, the declaration of structure *Date* is

```
Date dateObject, dateArray[4];
```

3.1.2 Accessing Members of Structures

To access an individual member of a struct variable, you can use the ***member access operators***, the *dot operator* (.) and the *arrow operator* (->). For example, to output member *day* of structure *dateObject* uses the following statements

```
Date dateObject;
cout << dateObject.day;
```

Assume that the *dataPtr* pointer has been declared to point to a *Date* object, and that the address of structure *dateObject* has been assigned to *datePtr*. To output member *day* of structure *dateObject* with pointer *datePtr*, uses the following statements

```
Date* datePtr = &dateObject;
cout << datePtr -> day;
```

Implementing a User-Defined Type with a Struct

The program defines a single *Date* structure called *today* and uses the dot operator to initialize the structure members with the values 18 for *day*, 11 for *month*, and 2010 for *year*. The program then adds a number to member *year* by calling function *add_year*, and prints the date in the English standard format.

Example 3-1: A user-defined type *Date* with a struct.

```
//-------------------------------------------------------------------------------
//File: example3_1.cpp
//This program defines a structured type Date and declares a structured variable today.
//The program adds a number to the year of today and prints the date by calling two functions
//that are defined outside of the structured type.
//-------------------------------------------------------------------------------
1   #include <iostream>
2   using namespace std;
3   //Defining a structure
```

```
4      struct Date {
5          int day;
6          int month;
7          int year;
8      };
9      //Defining functions for manipulating members of the structure
10     void init_date(Date& date, int d, int m, int y);
11     void add_year(Date& date, int n);
12     void print(Date& date);
13
14     int main()
15     {
16         Date today;                     //declaring a structure type
17
18         init_date(today, 18, 11, 2010); //set members to valid values
19
20         cout<<" The date of today is:";
21         print(today);                   //print today
22         cout << endl;
23
24         add_year(today, 2);             //add 2 to year
25         cout << " The date of today is:";
26         print(today);                   //print today
27         return 0;
28     }
29
30     //Initialize a date
31     void init_date (Date& date, int d, int m, int y)
32     {
33         date.day = d;    date.month = m;   date.year = y;
34     }
35     //Add a number to member year
36     void add_year(Date& date, int n)
37     {
38         date.year += n;
39     }
40     //Print a date in the English standard format
41     void print(Date& date)
42     {
43         cout << date.month << "-" << date.day << "-" << date.year;
44     }
```

Result:

```
1    The date of today is:11-18-2010
2    The date of today is:11-18-2012
```

In this example, the *Date* structured type contains three members of integer types, namely, *day*, *month* and *year*. When you declare a *today* variable of type *Date*, *today* does not represent just one data value of the member for type *Date*, it represents an entire collection of characters. Each of the members in *today* can be accessed individually, as the statement in Line 38 or 43 described. You may notice that three functions in Lines 31, 36 and 41 are defined for manipulating the members of *Date*.

Think It Over

Why are the functions passed by a reference parameter of type *Date* in Example 3-1?

3.1.3 Structures with Member Functions

In Example 3-1, we used the structure to define a date with variables and a set of functions for manipulating variables on this type. However, there is no explicit connection between the data type and these functions. Such a connection can be established by declaring the functions as members inside the structure. Thus, Example 3-1 can be changed into the following form.

```
1   #include <iostream>
2   using namespace std;
3   //Defining a structure
4   struct Date {
5       int day;
6       int month;
7       int year;
8
9       void init_date(int d, int m, int y);
10      void add_year(int n);
11      void print();
12  };
13
14  int main()
15  {
16      Date today;                        //declaring a structure type
17
18      today.init_date(18, 11, 2010);     //set members to valid values
19
20      cout << " The date of today is:";
21      today.print();                     //print today
22      cout<<endl;
23
24      today.add_year(2);                 //add 2 to year
25      cout << " The date of today is:";
26      today.print();                     //print today
27      return 0;
```

```
28    }
29    //Initialize a date
30    void Date::init_date (int d, int m, int y)
31    {
32        day = d;   month = m;   year = y;
33    }
34    //Add a number to member year
35    void Date::add_year(int n)
36    {
37        year += n;
38    }
39    //Print a date in the English standard format
40    void Date::print()
41    {
42        cout << month << "-" << day << "-" << year;
43    }
```

Result:

```
1    The date of today is:11-18-2010
2    The date of today is:11-18-2012
```

The result of this program is the same as the one of Example 3-1. However, the functions defined in Lines 9, 10 and 11 are placed inside the *Date* structure as members. These functions are accessed using a member access operator (.), as mentioned in Lines 18 and 21.

3.2 Data Abstraction and Classes

3.2.1 Data Abstraction

As the developed software becomes more complex, we design methods and data structures in parallel. We progress from the logical or abstract data structure envisioned at the top level through the refinement process until we reach the concrete coding.

Let's take the example of the TV which you can turn on and off, change the channel and adjust the volume. You know how to use a TV. But you do not know its internal detail, that is, you do not know how it receives signals over the air or through a cable, how it translates them and finally displays them on the screen.

Separating the design detail (that is, how the TV's works) from its use is called abstraction. In the words, abstraction focuses on what a TV does but not on how it works.

> **Data abstraction** is a process of separating the logical properties from the implementation detail. It is an OO methodology.
>
> 数据抽象是将逻辑属性与实现细节分离的过程，是一种面向对象的方法。

In this example, turning on a TV is a logical property. The internal process of turning on constitutes the implementation detail. The user is not interested in this process.

Abstraction can also be applied to data. We presented a *Date* structured type in Example 3-1. Now, let us illustrate data abstraction using the *Date* type. Definition of a *Date* type is as follows:

(1) Domain

Each *Date* value is a date in the form of year, month and day, that is, the *Date* type has three member variables, i.e. *year*, *month* and *day*.

(2) Operations

• Initialize a date (e.g. *void init_date(int d, int m, int y)*);

• Add a number to the year (e.g. *void add_year(int n)*);

• Print a date in the English standard format (e.g. *void print()*).

The actual implementation of these functions manipulating the member variables is separated.

3.2.2 Defining Classes

Chapter 1 introduced the programming methodology called object-oriented programming (OOP). In OOP, the first step is to identify the components (called objects) by using the principle of data abstraction. An object is the variable of a class which combines data and the operations on that data in a single unit. The aim of the C++ class concept is to provide the programmer with a tool for creating new types that can be used as conveniently as the built-in types.

A *class* is a user-defined type. Classes enable the programmer to model objects that have *attributes* (represented as ***data members***) and *behaviors* or *operations* (represented as ***member functions***). Types containing data members and member functions are defined in C++ using the ***class*** keyword.

Member functions are sometimes called ***methods*** in other object-oriented programming languages and are invoked in response to a message sent to an object. A message corresponds to a member-function call sent from one object to another or sent from a function to an object.

> A **class** is a user-defined type. A class is a set or collection of abstracted objects that share common characteristics. It consists of both data values and operations on those values.
> 类是一种用户定义的类型。类是具有共同特征的抽象对象的集合。它由数值和对这些数值的操作组成。

Classes are generally declared using the ***class*** keyword, with the following format:

    class class_name {
    access_specifier1:
        //members;
    access_specifier2:
        //members;
    };

where *class_name* is a valid identifier for the class. The body of the declaration within braces can contain members, which can be either data or function declaration and optionally access specifiers (see §3.4).

Like a structure, the body of the class is delimited by braces ({}) and terminated by a semicolon (;). (Do not miss the semicolon. Remember, data constructs, such as structures and classes, end by using a semicolon, whereas control constructs, such as functions and loops, do not.)

> The components of a class are called the **members of the class**. The members of the class may be either data type or functions. Data in the class are called **data members**. Functions declared within the class definition are called **member functions**.
> 类的组成部分称为类的成员。类的成员可以是数据类型或函数。类中的数据称为**数据成员**。类定义中声明的函数称为**成员函数**。

For example,

```
class Date{
    private:                    //access control specifier
        int day;                //1-31
        int month;              //1-12
        int year;               //1-9999
    public:                     //access control specifier
        void init_date (int d, int m, int y);    //initialize
        void add_year ( int n);                  //add n years
        void print();                            //print a date
};
```

In this example, integer variables *day*, *month* and *year* are the *data members* of the *Date* class; functions *init_date*, *add_year* and *print* are its *member functions*; the ***private*** and ***public*** keywords are the access specifiers.

3.2.3 Defining Objects

Once a class has been defined, the class name can be used to declare objects of that class.

> An **object** is an instance of a class or a variable of a class.
> 对象是一个类的实例或一个类的变量。

An object does not exist until an instance of the class has been created; the class is just a definition. When the object is physically created, the space for that object is allocated in the memory. It is possible to have multiple objects created from one class.

Objects are declared like variables of other types. The declaration

```
Date today, tomorrow;
Date *yesterday;
```

creates three objects of the *Date* class: *yesterday*, *today* and *tomorrow*. Each object has its own copies of *day*, *month* and *year*, the private data members of the class.

3.2.4 Accessing Member Functions

Like a structure variable, members of an object are accessed using the ***member access operators***—the *dot operator* (.) and the *arrow operator* (->). For example,

```
today.init_date(18,11,2010);
tomorrow.init_date(19,11,2010);
day->init_date(23, 8, 2003);
```

At a given moment during program execution, data members *day*, *month* and *year* of the *today* object are assigned with values 18, 11 and 2010 by invoking function *init_date*, whilst *tomorrow*'s data members are assigned with values 19, 11 and 2010.

Figure 3-1 shows a visual image of the class objects *today* and *tomorrow*.

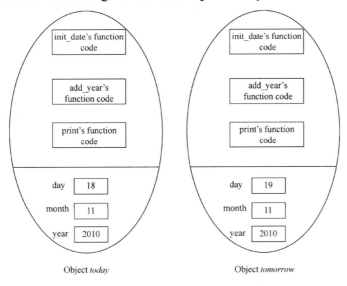

Figure 3-1　Conceptual view of two objects

Implementing an Abstract Data Type with a Class

Data abstraction is defined as a process of separating the logical properties of the data from its implementation. Every data type consists of a set of values (data members) along with a collection of operations (member functions) on those values. For example, the data members (e.g. *year*, *month* and *day*) of class *Date* and its basic operations are the logical properties; the algorithms to perform these operations are the implementation details of *Date*.

> An **abstract data type (ADT)**—A data type that separates the logical properties from the implementation details.
> 抽象数据类型（ADT）——一种将逻辑属性与实现细节分离的数据类型。

Example 3-2: An abstract data type—class *Date*.

```
//---------------------------------------------------------------
//File: example3_2.cpp
//This program defines a Date class type and creates the Date objects.
//---------------------------------------------------------------
1    #include <iostream>
2    using namespace std;
3    class Date {                    //class definition
4         int day;
5         int month;
6         int year;
7    public:
8         //Initialize a date
9         void init_date ( int d, int m, int y)
10        {
11             day = d;   month = m;   year = y;
12        }
13        //Add a number to member year
14        void add_year ( int n)
15        {
16             year += n;
17        }
18        //Print a date in the English standard format
19        void print ()
20        {
21             cout << month << "-" << day << "-" << year;
22        }
23   };
24
25   int main()
26   {
27        Date today, tomorrow;        //create two objects
28
29        today. init_date(18, 11, 2010);   //set members to valid values
```

```
30        tomorrow.init_date(19, 11, 2010);
31
32        cout << " The date of today is:";
33        today.print();
34        cout << endl;
35
36        cout << " The date of tomorrow is:";
37        tomorrow.print();
38        cout << endl;
39
40        today.add_year(2);      //add a number to year
41        tomorrow.add_year(2);
42
43        cout <<" The new date of today is:";
44        today.print();
45        cout << endl;
46        cout <<" The new date of today is:";
47        tomorrow.print();
48        cout <<endl;
49        return 0;
50  }
```

Result:

```
1  The date of today is:11-18-2010
2  The date of tomorrow is:11-19-2010
3  The new date of today is:11-18-2012
4  The new date of tomorrow is:11-19-2012
```

In a member function, member names can be used without explicit reference to an object for which the function was invoked. For example, function *init_date* is invoked by objects *today* and *tomorrow* respectively.

3.2.5 In-Class Member Function Definition

A member function defined within the class definition—rather than simply declared there—is taken to be an inline member function. That is, in-class definition of member functions is for small, frequently-used functions. Consider

```
1  //Date.cpp
2  class Date {
3  public:
4      void init_date(int, int, int);
5      void add_year(int n)
6      {  year += n;  }
7      void print();
8  private:
```

```
9     int day, month, year;
10 };
```

The statements in Lines 5 and 6 represent that the member function *add_year* is defined inside the class. The other functions are defined outside the class. This is perfectly good C++ code because a member function declared within a class can refer to every member of the class as if the class is completely defined before the member function bodies were considered. However, this can confuse human readers.

Consequently, we usually either place the data first or define the inline member functions after the class itself. For example,

```
1    //Date.cpp
2    class Date {
3    public:
4        void init_date(int, int, int);
5        void add_year(int );
6        void print();
7    private:
8        int day, month, year;
9    };
10
11   inline void Date::add_year(int n)
12   { year += n; }
```

Member Functions Defined Outside the Class

We have seen member functions that were defined inside the class definition in Example 3-2. This does not need to always be the case. The statement in Line 11 of the example above shows a member function *add_year* that is not defined within the *Date* class definition. It is only *declared* inside the class, with the statement

```
void add_year( int );
```

This tells the compiler that this function is a member of the class but that it will be defined outside the class declaration, someplace else in the listing.

In this situation, the member function is defined in the following syntax:

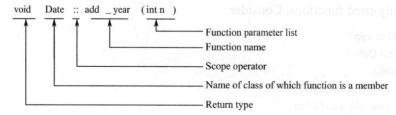

The *Date* class in Example 3-2 can also be defined as follows:

```
1    //Date.cpp
2    class Date {
```

```
3   public:
4       void init_date(int, int, int);
5       void add_year(int);
6       void print();
7   private:
8       int day, month, year;
9   };
10  void Date::init_date(int d, int m, int y)
11  {   day = d; month = m; year = y;   }
12  void Date::add_year(int n)
13  {   year += n;   }
14  void Date:: print()
15  {   cout << month << "-" <<day << "-" << year;   }
```

3.2.6 File Structure of an Abstract Data Type

Normally, the *Date* class declaration serves as the specification of *Date*. It is described in a specification file which is called a header file (or interface file).

> An **abstract data type (ADT)** consists of two parts: a **specification** and an **implementation**. The specification describes the behavior of the data type without reference to its implementation. The implementation creates an abstraction barrier by hiding the concrete data representation as well as the code for the operations.
> 抽象数据类型（ADT）由两部分组成：**说明和实现**。说明在不引用其实现的情况下描述数据类型的行为，而实现通过隐藏具体的数据表示及操作的代码来创建抽象屏障。

Example 3-3: File structure of the *Date* abstract data type.

```
//-------------------------------------------------------------------
//File: Date.h (specification file)
//This program describes the definition of a Date abstract date type.
//-------------------------------------------------------------------
1   class Date {
2   public:
3       void init_date(int, int, int);
4       void add_year(int);
5       void print();
6   private:
7       int day;
8       int month;
9       int year;
10  };
```

The specification file (.h) for the *Date* class contains only the class declaration. The implementation file (.cpp) must provide the definitions of all class member functions.

```
//-----------------------------------------------------------------------
//File: Date.cpp (implementation file)
//The program presents the implementation detail of the Date class.
//-----------------------------------------------------------------------
1   #include "Date.h"                        //referring the declaration of the class Date
2   #include <iostream>
3
4   using namespace std;
5
6   //Initialize a date
7   void Date::init_date(int d, int m, int y)
8   {
9       day = d; month = m; year = y;
10  }
11  //Add a number to member year
12  void Date::add_year( int n)
13  {
14      year += n;
15  }
16  //Print a date in the English standard format
17  void Date::print()
18  {
19      cout << month << "-" << day << "-" << year;
20  }
```

The most important new thing in this code is the operator of scope (::, two colons) included in the definition of three member functions *init_date, add_year* and *print*. With such operations, the members of a class can be defined outside the class.

The scope operator (::) specifies the class to which the declared member belongs, granting exactly the same scope properties as if this function definition was directly included within the class definition (compared with Example 3-2).

For example, in the function *init_date* of the previous code, we have been able to use the variables *day*, *month* and *year*, which are private members of class *Date*, which means they are only accessible to other members of their class.

You should have a user file (.cpp) to use the *Date* class.

```
//-----------------------------------------------------------------------
//File: example3_3.cpp (user file)
//This program creates the Date objects and accesses their members.
//-----------------------------------------------------------------------
1   #include "Date.h"                        //referring the declaration of the class
2   #include <iostream>
3
4   using namespace std;
5   int main()
```

```
6      {
7              Date today, tomorrow;              //create two objects
8
9              today. init_date(18, 11, 2010);    //set members to valid values
10             tomorrow.init_date(19, 11, 2010);
11
12             cout << " The date of today is:";
13             today.print();
14             cout << endl;
15
16             cout << " The date of tomorrow is:";
17             tomorrow.print();
18             cout << endl;
19
20             today.add_year(2);                 //add a number to year
21             tomorrow.add_year(2);
22
23             cout << " The new date of today is:";
24             today.print();
25             cout << endl;
26             cout << " The new date of today is:";
27             tomorrow.print();
28             cout << endl;
29             return 0;
30     }
```

Result:

```
1    The date of today is:11-18-2010
2    The date of tomorrow is:11-19-2010
3    The new date of today is:11-18-2012
4    The new date of tomorrow is:11-19-2012
```

3.3 Information Hiding

The previous section defined the *Date* class to implement the data operations in a program. We then wrote a program (Date.cpp) that used the *Date* class. In fact, we combined the *Date* class with the function definitions to implement the operations and the *main* function so as to complete the program. That is, the specification and implementation details of the *Date* class were directly incorporated into the program.

Is it a good practice to include the specification and implementation detail of a class in the program? Definitely not. There are several reasons for not doing so.

Suppose the definition of the class and the definitions of the member functions are directly included in the user's program (see Example 3-2). The user then has direct access to the definition of the class and the definitions of the member functions. Therefore, the user can modify the operations in any way the user pleases. The user can also modify the data members of an object in any way the user pleases. Thus, in this sense, *the private data members of an object are no longer private to the object.*

If several programmers use the same object in a project with direct access to the internal parts of the object, there is no guarantee that every programmer will use the same object in exactly the same way. Thus, we must hide the implementation details. *The user should know only what the object does, not how it does.*

Hiding the implementation details frees the user from having to fit this extra piece of code in the program. Also, by hiding the details, we can ensure that an object will be used in exactly the same way throughout the project. Furthermore, once an object has been written, debugged, and tested properly, it becomes (and remains) error-free.

To implement Date in a program, the user must declare objects of class Date, and know what operations are allowed and what the operations do. To do so, the user must have access to the specification details. As the user is not concerned with the implementation details, we must put those details in a separate file (e.g. Date.cpp in Example 3-3).

Additionally, we must free the user from having to include them directly in the program because the specification details can be too long.

For example, in the example3_3.cpp program of Example 3-3, the user can do:

```
#include "Date.h"
```

Notice that the user must be able to look at the specification details so that he or she can correctly call the functions, and so forth.

Think These Over
1. Why do we need to use data abstraction in OOP?
2. What is an abstract data type?
3. What is the purpose of information hiding?

3.4 Access Control

All are very similar to the declaration on data structures, except that we can not only include functions and members, but also this new thing called *access specifier*. An access specifier is one of the following three keywords: *private*, *public* or *protected*. These specifiers modify the access rights that the members following them acquire:

private members of a class are accessible only to other members within the same class

• 66 •

and friends (see §4.8);

protected members are accessible to members of their own class and friends, but also to members of their derived classes (see §6.5.1);

public members are accessible from anywhere where the object is visible.

The body of the class contains two unfamiliar keywords: private and public. What is its purpose?

A key feature of object-oriented programming is ***information (data) hiding***. This term does not refer to the activities of particularly paranoid programmers; rather it means that data is concealed within a class so that it cannot be accessed mistakenly by functions outside the class. The primary mechanism for hiding data is to put it in a class and make it private. ***Private*** data or functions can only be accessed from within the class. ***Public*** data or functions, on the other hand, are accessible from outside the class.

The declaration of the *Date* class in the previous section provides a set of functions for manipulating a date. However, it does not specify that those functions should be the only ones to depend directly on *Date*'s representation and the only ones to access directly to the *Date* objects. This restriction can be expressed by using a ***class*** instead of a ***struct***:

```
class Date{
private:
    int day;                              //1-31
    int month;                            //1-12
    int year;                             //1-9999
public:                                   //access control specifier
    void init_date(int d, int m, int y);  //initialize
    void add_year(int n);                 //add n years
    void print();                         //print a date
};
```

The public specifier separates the class body into two parts. The names in the first part with the private keyword, including data members *day*, *month* and *year*, can be used only by member functions. The second part with the public keyword, including member functions *init_date*, *add_year* and *print*, constitutes the public interface to objects of the class. A *struct* is simply a *class* whose members are public by default; member functions can be defined and used exactly as before.

The private part of the members can be used only by member functions. In Example 3-2, the member function definition is:

```
void init_date(int d, int m, int y)
{    day = d;   month = m;   year = y;   }
```

Here, *day*, *month* and *year* are all data members of class *Date*. The function *init_date* is a member function of class *Date*. The members *day*, *month* and *year* can be accessed by function *init_date*.

However, non-member functions are forbidden from using private members. For example:

```
int main()
{
    Date d;
    d.year -= 200;      //error: year is private
}
```

Private data or functions can only be accessed from within the class. Public data or functions, on the other hand, are accessible from outside the class. By default, all members of a class declared with the class keyword have private access for all its members. Therefore, any member that is declared before the class specifier automatically has private access.

Notice that the main function is not a member function of the *Date* class.
私有数据或函数只能在类内被访问。公有数据或函数可以在类外部被访问。在默认情况下，用 Class 关键字声明的类的所有成员对其所有成员具有私有访问权。因此，在类说明符之前声明的任何成员都具有私有访问权限。

请注意，main 函数不是 Date 类的成员函数。

The ***public*** part of the members can be used by member functions of the class or other functions. For example,

```
int main()
{
    Date d;
    d.init_date (2, 6, 1999)    //ok: init_date is public
}
```

Structures vs Classes

A *struct* in C is defined as a fixed collection of components, wherein the components can be of different types. This definition of components in a *struct* includes only member variables. However, a C++ *struct* is very similar to a *class*. By definition, a *struct* is a class in which members are by default public; that is,

```
struct Date {…};
```

is simple shorthand for

```
class Date { public: …};
```

The access specifier ***private*:** can be used to indicate that the members following are private, just as ***public*:** says that the members following are public.

A **class** resembles a **struct** with just one difference—all struct members are public by default, but all class members are private by default.

类与**结构体**只有一点不同——在默认情况下，所有结构成员都是公有的，但所有类成员都是私有的。

Except for the different names, the declarations shown in Figure 3-2 are equivalent.

```
struct Date{                        class Date{
    private:                            int day, month, year;
        int day, month, year;       public:
    public:                             void add_year(int n);
        void add_year(int n);       };
};
```

Figure 3-2　Declarations of struct *Date* and class *Date*

3.5　Constructors

The use of the function, such as *init_date,* to provide initialization for class objects is inelegant. Because it is nowhere stated that an object must be initialized, a programmer may forget to do so—or do so twice (often with equally disastrous results).

A better approach is to allow the programmer to declare a function with the explicit purpose of initializing objects. Since such a function constructs values of a given type, it is called a ***constructor***.

> A **constructor** is a member function that is implicitly invoked whenever a class object is created.
> 构造函数是每当创建类对象时隐式调用的成员函数。

3.5.1　Definition of Constructors

A constructor is recognized by having the same name as the class itself. It serves the purpose of initializing objects. For example,

```
//In the Date.h file, a constructor is added
class Date {
public:
    Date(int, int, int);        //constructor with parameters
    void add_year(int);
    void print();
private:
    int day, month, year;
```

```
    };
    //Definition of the constructor outside the class in the Date.cpp file
    Date::Date(int d, int m, int y)
    {
        day = d;   month = m;   year = y;
    }
```

Initializing Objects with Constructors

When an object of the *Date* type is first created, we want its data to be initialized to 18, 11, 2010, for example,

```
    Date today(18, 11, 2010);
```

This statement means that the constructor of class *Date* is automatically called, and data members *day*, *month* and *year* of the *Date* are initialized with values 18, 11 and 2010 when object *today* is created.

When a class has a constructor, all objects of that class will be initialized. If the constructor requires arguments, these arguments must be supplied. For example,

```
    int main()
    {
        Date today(18, 11, 2010);      //ok
        Date tomorrow(19, 11, 2010);   //ok
        Date my_birthday;              //error: initializer missing
        Date xmax(25, 12);             //error: 3rd argument missing
    }
```

3.5.2 Overloading Constructors

To guarantee that the members of a class are initialized, you use constructors. There are two types of constructors: with parameters and without parameters. The constructor without parameters is called the ***default constructor***.

It is often nice to provide several ways of initializing a class object. This can be done by providing several constructors. The constructors obey the same overloading rules as other functions do. As long as the constructors sufficiently differ in their parameter types, the compiler can select the correct one for each use.

For example,

```
//----------------------------------------------------------------------
//This program overloads the constructors of class Date.
//----------------------------------------------------------------------
1   class Date{
2   public:
3       Date(int, int, int);      //constructor with three integer parameters
4       Date(int, int);           //constructor with two integer parameters
```

```
5      Date(int);                    //constructor with one integer parameter
6      Date(const *char);            //constructor with a char pointer parameter
7   private:
8      int day, month, year;
9   };
10
11  int main()
12  {
13     Date today(18, 11, 2010);
14     Date xmas(25, 12);
15     Date d("18-11-2010");
16     return 0;
17  }
```

3.5.3 Constructors with Default Parameters

A constructor can also have default parameters. In such class, the rules for declaring formal parameters are the same as those for declaring default parameters in a function. Moreover, actual parameters to a constructor with default parameters are passed according to the rules for functions with default parameters.

> A constructor is called a **default constructor** in the following three situations:
> (1) The constructor isn't defined in the class;
> (2) The constructor is without parameters;
> (3) The constructor is with default parameters;
> 构造函数在下列三种情况下称为**默认构造函数**：
> (1) 类中没有定义构造函数；
> (2) 构造函数没有参数；
> (3) 构造函数带有默认参数。

Example 3-4: An example of the *Date* class with the default constructor.

```
//---------------------------------------------------------
//File: Date.h (specification file)
//This program defines the Date class.
//---------------------------------------------------------
1   class Date {
2   public:
3      //Declaration of the constructor with default parameters
4      Date(int = 1, int = 1, int = 2000);
5      void add_year (int);
6      void print();
7
8   private:
9      int day;
10     int month;
11     int year;
```

· 71 ·

```
12   };
```

In the implementation file, the definition of the constructor is the same as the definition of the constructors with parameters.

```
//------------------------------------------------------------------
//File: Date.cpp (implementation file)
//This program defines the implementation detail of the Date class.
//------------------------------------------------------------------
1   #include "Date.h"         //referring the declaration of the class
2   #include <iostream>
3   using namespace std;
4
5   //Definition of a constructor
6   Date::Date(int d, int m, int y)
7   {
8        day = d;   month = m;   year = y;
9   }
10  //Add a number to member year
11  void Date::add_year(int n)
12  {
13       year += n;
14  }
15  //Print a date in the English standard format
16  void Date::print()
17  {
18       cout << month << "-" << day << "-" << year;
19  }
```

In the user file, you can declare *Date* objects with zero, one, two, or three arguments, as follows:

```
//------------------------------------------------------------------
//File: example3_4.cpp (user file)
//This program creates Date objects with different parameter lists.
//------------------------------------------------------------------
1   void main()
2   {
3        Date day;
4        Date myday(23);
5        Date xmas(25, 12);
6        Date today(18, 11, 2010);
7   }
```

In Line 3, the members of *day* are initialized to be: *day* = 1, *month* = 1, *year* = 2000. In Line 4, the members of *myday* are initialized to be: *day* = 23, *month* = 1, *year* = 2000. In Line 5, the members of *xmas* are initialized to be: *day* = 25, *month* = 12, *year* = 2000. In Line 6, the members of *today* are initialized to be: *day* = 18, *month* = 11, *year* = 2010.

Sometimes we also want to define the variables of type *Date* without initializing them, for example,

```
class Date {
public:
    Date()              //constructor without parameters
    { day = 1;   month = 1;   year =2015; }
private:
    int day, month, year;
};
```

When the objects are created as follows,

```
Date today, someday;
```

there is no constructor fitting in the statement above, but our definitions worked just fine. How could they work without a constructor? Because an implicit no-argument constructor is built into the program automatically by the compiler. It is the default constructor that created the objects, even though we did not define it in the class.

If you declare an object and want the default constructor to be executed, the declaration

```
Date day;        //ok: invoking the default constructor
```

is legal. However, if you declare the *day* object in the following format:

```
Data day();      //error
```

it is illegal.

A constructor is similar to a member function, but with the following differences:
- No return type.
- No *return* statement.
- The name of the constructor is the same as the name of the class.

• A class can have more than one constructor. However, all constructors of a class must have the same name.
• If a class has more than one constructor, the constructor must have different formal parameter lists. That is, either they have a different number of the formal parameters or, if the number of formal parameters is the same, the data type of the formal parameters must differ in at least one position.
• Constructors is executed automatically when a class object is created.
• Which constructor to execute depends on the types of values passed to the class object when the class object is declared.

构造函数类似于成员函数，但具有以下不同：
- 没有返回类型。
- 没有 return 语句。

- 构造函数名与类名相同。
- 一个类可以有不止一个构造函数。但是，类的所有构造函数都具有相同的名称。
- 如果一个类有不止一个构造函数，则构造函数必须具有不同的形参列表，也就是说，它们要么具有不同数量的形参，如果形式参数的数量相同，那么形参的数据类型必须在至少一个位置上不同。
- 当创建类对象时，构造函数自动执行。
- 执行哪个构造函数取决于声明类对象时传递给类对象的值的类型。

3.6 Destructors

3.6.1 Definition of Destructors

A constructor initializes an object. In other words, it creates the environment in which the member functions operate. Sometimes, creating that environment involves acquiring a resource, such as a file, a lock, or some memory, that must be released after use. Thus, some classes need a function that is guaranteed to be invoked when an object is destroyed in a manner similar to the way a constructor is guaranteed to be invoked when an object is created.

Destructors are usually used to deallocate memory and do other cleanup for a class object and its class members when the object is destroyed. A destructor is called for a class object when that object passes out of scope or is explicitly deleted.

> A **destructor** is a member function that is used to clean up and release resources.
> 析构函数是用于清理和释放资源的成员函数。

A destructor is a member function with the same name as its class prefixed by a ~ (tilde). For example:

```
1    //define a destructor of class Date
2    class Date {
3    public:
4        Date(int, int, int);              //constructor
5        //other member functions
6        ~Date();                          //destructor
7    private:
8        //data members
9    };
```

A destructor takes no arguments and has no return type. Its address cannot be taken. Destructors cannot be declared as const, volatile, const volatile or static. A destructor can be declared as virtual or pure virtual (see §7.2.4).

If no user-defined destructor exists for a class, the compiler implicitly declares a destructor. This implicitly-declared destructor is an inline public member of its class.

The compiler will implicitly define an implicitly declared destructor when the compiler uses the destructor to destroy an object of the destructor's class type. Suppose a *Date* class has an implicitly declared destructor. The following is equivalent to the function the compiler would implicitly define for *Date*:

	Date::~Date() { }	//definition outside the *Date* class
or		
	~Date() { }	//definition inside the *Date* class

A destructor is a member function, but with the following differences:

- No return type.
- No *return* statement.
- No parameters.
- The name of a destructor is the same as the name of the class prefixed by a ~ (tilde).
- A class can only have one destructor—the destructor cannot be overloaded.
- The destructor is executed automatically when a class object goes out of the scope.

析构函数是一个成员函数，但具有以下不同：

- 没有返回类型。
- 没有 return 语句。
- 没有参数。
- 析构函数名与类名相同，但有一个~前缀。
- 一个类只能有一个析构函数，即析构函数不能被重载。
- 当类对象离开它的作用域范围时，析构函数将被自动执行。

3.6.2 UML Diagram for Classes

The Unified Modelling Language (UML) is now the most widely used graphical representation scheme for modeling object-oriented system. It has indeed unified the various popular schemes. Those who design systems use the language (in the form of the diagram) to model their system, as we do throughout this book.

In the UML, each class is modeled in a class diagram as a rectangle with three compartments. Figure 3-3 presents a UML class diagram for the *Date* class in the example above. The top compartment describes the class name. The middle compartment contains the data members *day*, *month* and *year*. The bottom compartment presents the member functions *add_year* and *print*. Of course, the constructor and destructor of the *Date* class are also included within this compartment. The plus sign (+) in front of member names indicates that these members are public. The minus sign (-) indicates private members.

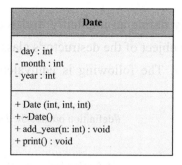

Figure 3-3 UML class diagram for class *Date*

3.6.3 The Order of Constructor and Destructor Calls

Constructors and destructor are called automatically by the compiler. The order in which these function calls are made depends on the order in which their execution enters and leaves the scopes in which objects are instantiated. Generally, the destructor calls are made in the reverse order of the corresponding constructor calls.

Example 3-5: The order of constructor and destructor calls.

```
//---------------------------------------------------------------
//File: TestOrder.h
//This program presents the specification of class TestOrder.
//---------------------------------------------------------------
1    class TestOrder {
2    public:
3        TestOrder(int);        //constructor
4        ~TestOrder();          //destructor
5    private:
6        int data;
7    };
//---------------------------------------------------------------
//File: TestOrder.cpp
//This program presents the implementation of class TestOrder.
//---------------------------------------------------------------
1    #include "TestOrder.h"
2    #include <iostream>
3    using namespace std;
4
5    TestOrder::TestOrder(int d)
6    {
7        data = d;
8        cout << "Object " << data << " is constructed\n";
9    }
10
11   TestOrder::~TestOrder()
12   {
```

```
13          cout << "Object " << data << " is destroyed\n";
14      }
```
//--
//File: **example3_5.cpp**
//This program demonstrates the order of constructor and destructor calls.
//--
```
1    #include "TestOrder.h"
2    #include <iostream>
3    using namespace std;
4
5    //a sub-function using the class object
6    void foo()
7    {
8        TestOrder furthObj(200);        //local object
9    }
10
11   TestOrder firstObj(1);              //global object
12
13   int main()
14   {
15       TestOrder secondObj(100);
16
17       for (int i = 2; i <= 4; i++)
18       {
19           TestOrder loopObj(i);       //local objects
20       }
21
22       foo();
23
24       return 0;
25   }
```

Result:

```
1    Object 1 is constructed
2    Object 100 is constructed
3    Object 2 is constructed
4    Object 2 is destroyed
5    Object 3 is constructed
6    Object 3 is destroyed
7    Object 4 is constructed
8    Object 4 is destroyed
9    Object 200 is constructed
10   Object 200 is destroyed
11   Object 100 is destroyed
12   Object 1 is destroyed
```

The program above demonstrates the order in which constructors and destructors are called for objects of class *TestOrder* in several scopes.

The program defines the object *firstObj* in Line 11 (Example 3-5.cpp) in the global scope. Its constructor is called as the program begins execution and its destructor is called at program termination after all other objects are destroyed.

The main function declares four local objects. One is *secondObj* in Line 15. The other three objects named *loopObj* in Line 19 are in the *for* loop body. The constructor for *secondObj* is called when execution reaches the point where this object is declared. The destructor for *secondObj* is called when the end of the *main* function is reached. The constructors for *loopObj* are called when execution enters the loop body and the destructors for *loopObj* are called when execution leaves the loop body.

The *foo* function declares object *furthObj*. The constructor for *furthObj* is called when execution reaches the point where *furthObj* is declared. The destructor for *furthObj* is called when reaching at the end of the *foo* function.

Think These Over

1. What is the purpose of using constructors and destructors in a class?
2. What are the differences between constructors/destructors and member functions of the class?

3.7 Encapsulation

Encapsulation is one of the fundamental principles of object-oriented programming. It is the grouping of related ideas into one unit, which can thereafter be referred to by a single name. The process is combining data and functions into a single unit called ***class***.

Encapsulation means that all of the object's data is contained and hidden in the object and such data access is restricted to the members of that class, that is, the programmer cannot directly access the data. Data is only accessible through the functions existing inside the class.

Example 3-6: Design the *Counter* objects by using the concept of encapsulation.

```
//-----------------------------------------------------------------
//File: Counter.h
//The program presents the declaration of class Counter.
//-----------------------------------------------------------------
1    class Counter{
2    public:
3        //constructor with default parameter
4        Counter(int i = 0);
5        //interface to outside world
6        void add();
7        //interface to outside world
```

```
8       int getSum();
9   private:
10      //hidden data from outside world
11      int sum;
12  };
```
//---
//File: **Counter.cpp**
//The program presents the implementation of class *Counter* and tests it.
//---
```
1   #include "Counter.h"
2   #include <iostream>
3   using namespace std;
4
5   Counter::Counter(int i)
6   { sum = i; }
7   void Counter::add()
8   { sum++; }
9   int Counter::getSum()
10  { return sum; }
11
12  int main()
13  {
14      Counter c1;
15      c1.add();
16      c1.add();
17      c1.add();
18      cout << "The sum of the counter is " << c1.getSum() << endl;
19
20      Counter c2(20);
21      c2.add();
22      c2.add();
23      cout << "The sum of the counter is " << c2.getSum() << endl;
24
25      return 0;
26  }
```

Result:

```
1   The sum of the counter is 3
2   The sum of the counter is 22
```

The *Counter* class increments a number by one successively and returns the sum. The public member functions *add* and *getSum* are the interfaces to the outside world and a user needs to know them to use the class. The private member *sum* is something that is hidden from the outside world but is needed for proper class operation.

In a word, Data encapsulation leads to the important concept of data hiding. Data hiding is the implementation details of a class that are hidden from the user.

3.8 Case Study: A GradeBook Class

Now, let us take an example to demonstrate how to create a class and how to use it.

We design a grade book that helps a teacher to efficiently manage his/her student grades. A class, named *GradeBook*, has the following requirements:
- Store the attributes of a course, i.e. course name, teacher and hour.
- Display these attributes.
- Modify the attributes.

According to the statements above, the UML diagram of the *GradeBook* class is shown in Figure 3-4.

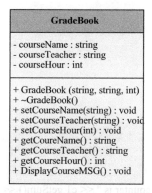

Figure 3-4   The UML diagram for the *GradeBook* class

Example 3-7. Demonstrate how to define a *GradeBook* class and use it.

```
//------------------------------------------------------------------------
//File: GradeBook.h
//The program presents the declaration of class GradeBook.
//------------------------------------------------------------------------
1   class GradeBook {
2   public:
3       //constructor initializes data members courseName, courseTeacher and courseHour
4       GradeBook(string coursename, string courseteacher, int coursehour );
5       //three set functions to reset the values of courseName, courseTeacher and courseHour
6       void setCourseName(string name );
7       void setCourseTeacher(string teachername );
8       void setCourseHour(int classhour);
9       //three get function to obtain the values of courseName, courseTeacher and courseHour
10      string getCourseName();
11      string getCourseTeacher();
12      int getCourseHour();
13      //display course messages and to the GradeBook user
```

```
14      void displayCouseMSG();
15      ~GradeBook();
16   private:
17      string courseName;        //course name
18      string courseTeacher;     //course teacher
19      int courseHour;           //course hour
20 };
```

First, we declare three private data members *courseName*, *courseTeacher* and *courseHour* in the *GradeBook*. These data members represent the attributes of the class. Then, three *set* member functions are declared to reset the attribute values, and three *get* functions are used to obtain the attribute values. Finally, a *displayCourseMSG* function is declared.

```
//-------------------------------------------------------------------------
//File: GradeBook.cpp
//The program presents the implementation of class GradeBook and test it.
//-------------------------------------------------------------------------
1  #include <iostream>
2  using namespace std;
3  #include "GradeBook.h"              //include definition of class GradeBook
4  GradeBook::GradeBook(string coursename, string courseteacher, int coursehour )
5  {
6      setCourseName(coursename );     //call set function to initialize courseName
7      setTeacherName(courseteacher);  //call set function to initialize courseTeacher
8      setClassHour(coursehour);       //call set function to initialize courseHour
9  }
10
11 void GradeBook::setCourseName(string name )
12 {
13     courseName = name;              //store the course name in the object
14 }
15 void GradeBook::setCourseTeacher(string teacher )
16 {
17     courseTeacher = teacher;        //store the course teacher in the object
18 }
19 void setCourseHour(int hour)
20 {
21     courseHour = hour;              //store the course teacher in the object
22 }
23
24 string getCourseName()
25 {
26     return courseName;              //return object's courseName
27 }
28 string getCourseTeacher()
29 {
30     return courseTeacher;           //return object's courseTeacher
```

```
31    }
32    int getCourseHour()
33    {
34        return courseHour;              //return object's courseHour
35    }
36
37    void displayMessage()
38    {
39        cout << "Welcome to the grade book for\n" << getCourseName() << "!" << endl;
40        cout << "Teacher: " << getCourseTeacher() << "; " << "Hour: " << getCourseHour() << endl;
41    }
42
43    GradeBook::~GradeBook()
44    {   cout << "Destroying an object\n";   }
45    int main()
46    {
47        //create two GradeBook objects
48        GradeBook gradeBook1("Object-Oriented Programming in C++", "Christina", 48);
49        gradeBook1.displayCourseMSG();
50
51        GradeBook gradeBook2("JAVA programming", "Ellen", 40 );
52        gradeBook2.displayCourseMSG();
53        return 0;
54    }
```

Result:

```
1   Welcome to the grade book for Object-Oriented Programming in C++
2   Teacher: Christina; Hour: 48
3   Welcome to the grade book for JAVA programming
4   Teacher: Ellen; Hour: 40
5   Destroying an object
6   Destroying an object
```

In the GradeBook.cpp file, we present the definitions of member functions outside the class. To test this class, we create two objects of the class, namely *gradeBook1* and *gradeBook2*, in the *main* function. The two objects access the *displayCourseMSG* member function to display the information of these two courses respectively.

Word Tips

access   *vt.*   存取，访问   assume   *vt.*   假定
aggregate   *vt./vi.* 聚合   bar   *vt.*   阻止
allocate   *vt./vi* 分配   call   *vt./vi.* 调用
arrow   *n.*   箭头   compiler *n.*   编译器

component	*n.* 组成部分	manipulation	*n.* 操作
conflict	*vi.* 冲突	memory	*n.* 记忆，内存
constitute	*vt.* 组成，构成	modify	*vt./vi.* 修改
construct	*vt.* 构造	operator	*n.* 运算符
constructor	*n.* 构造函数	overload	*vt.* 重载
debug	*vt.* 排除故障	parameter	*n.* 参数
declaration	*n.* 声明	pointer	*n.* 指针
declare	*vt./vi.* 声明	prefix	*n.* 前缀
default	*n.* 缺省，默认	previous	*adj.* 先前的
definition	*n.* 定义	prone	*adj.* 有可能的
delimit	*vt* 定界	recognize	*vt./vi.* 认为
disastrous	*adj.* 灾难性的	release	*vt.* 释放
equivalent	*adj.* 等价的	representation	*n.* 代表
execution	*n.* 执行	reserve	*vt.* 预留
explicit	*adj.* 明显的	respectively	*adv.* 分别，各自
format	*n.* 格式	scope	*n.* 作用域
function	*n.* 函数	semicolon	*n.* 分号
guarantee	*vt.* 保证	specify	*vt.* 详述，指定
inelegant	*adj.* 粗俗的	structure	*n.* 结构体
initialize	*vt.* 初始化	tag	*n.* 标志
invoke	*vt.* 调用	value	*n.* 值
loop	*n.* 循环	variable	*n.* 变量

Exercises

1. Mark the following statements as true (T) or false (F) and give reasons.

(1) Objects normally are not allowed to know how other objects are implemented.

(2) The class is also referred to as programmer-defined types.

(3) A member function defined outside of the class where it is declared does not have class scope.

(4) A member function with the same name as the class and preceded with a ~ is called the constructor function.

(5) Member functions cannot be overloaded.

(6) Operators "." and " ->" are used to access class members.

(7) Header files provide the complete definition of a class's implementation details.

(8) Each class definition is normally placed in a header file.

(9) The default access mode for members of a class is private.

(10) The labels public or private can be repeated in a class definition.

(11) Member functions labeled public can only be accessed from outside the class.

(12) Access to members of a struct can be defined as public or private.

(13) Like any function, constructors can return a value.

(14) Constructors cannot be overloaded.

(15) A constructor is a member function with the same names as the class.

(16) A class has only one constructor.

(17) There can only be one default constructor per class.

(18) The constructor can contain default arguments in the parameter list.

(19) A destructor is called when an object is removed from memory.

(20) Like a constructor, a destructor can take arguments.

(21) A destructor may return a value.

(22) The destructor can be overloaded.

(23) Constructors and destructors are called automatically.

(24) The programmer must always provide a constructor and destructor for a class.

(25) A set function should perform the validity examination.

(26) Information hiding is a process by which a class's data implementation is hidden from the user of the class.

2. Find the syntax errors in the definitions of the following classes.

```
(1)  class AA {
     public:
         void AA(int, int);
         int sum();
     private:
         int x = 0;
         int y;
     };

(2)  class BB {
     public:
         BB(int, int);
         print();
     private:
         int x, y;
     }

(3)  class CC {
     public:
         CC();
```

```
            CC(int, int);
        private:
            int x, y;
    };
    int main()
    {   CC c1(4);
        CC c2(4, 5);
        CC c3;
        cout<< c2.x<<c2.y<< endl;
        return 0; }
```

3. Consider the following declaration:

```
class Point {
public:
    Point();
    Point(int, int);
    void move(int, int);
    void print();
    ~Point();
private:
    int x, y;
};
and assume that the following statements are in a user program:
    Point point1;
    Point point2(10, 20);
```

(1) How many members does class *Point* have?

(2) How many *private* members does class *Point* have?

(3) How many constructors does class *Point* have?

(4) Write the definition of default constructor of class *Point* so that the *private* member variables are initialized to be 0.

(5) Write the definition of the constructor with arguments *xx* and *yy* so that the *private* member variables are initialized to be the values of *xx* and *yy*.

(6) Write the definition of member function *move* with arguments *newX* and *newY* so that *x* is reset to be *newX* and *y* is reset to be *newY*.

(7) Write the definition of member function *print* that outputs the values of *x* and *y*.

(8) Write the *main* function to test class *Point*.

4. Write the definition of a *Word* class that implements the functions of adding a meaning, getting a meaning, getting the number of word meanings and outputting all information of a word.

5. Write a *TimeDemo* class that includes the following properties:

(1) Three data members of *hour*, *minute* and *second*.

(2) Three functions to set three data members respectively.

(3) Time output in form of 12 hours or 24 hours, such as

12 hours:	9:45AM	3:10PM
24 hours:	9:45	15:10

(4) A constructor with default parameters (default time is 0:0:0).

6. Define a *Rectangle* class with data members *length* and *width*, each of which is 1 by default.

(1) Two public member functions to calculate the perimeter and area of the rectangle.

(2) The public *set* functions change the values of *length* and *width* and verify that *length* and *width* are integers larger than 0 and less than 50 respectively.

(3) The public *get* functions return the values of *length* and *width*.

Chapter 4 Advance of Classes and Objects

—*Further Definition of Class Members and Objects*

> *Those types are not "abstract",*
> *they are as real as* int *and* float.
> —*Doug McIlroy*

Objectives

- To specify *const* members and *const* objects
- To be able to use *static* members
- To understand the use of the *this* pointer
- To be able to create objects of other classes as class members
- To understand the purpose of *friend* functions and *friend* classes
- To understand how to use shallow copy and deep copy

4.1 Constant Member Functions and Constant Objects

The *Date* class defined so far provides member functions in order to give a *Date* object values and change it. Unfortunately, we did not provide a way of examining the value of a *Date*. This problem can easily be remedied by adding functions for reading the day, month, and year.

Example 4-1: Definition of a *Date* class with constant member functions.

```
//-------------------------------------------------------------------------
//File: Date.h (specification file)
//This program defines a Date class with constant member functions.
//-------------------------------------------------------------------------
1    class Date {
2    public:
3        Date(int = 1, int = 1, int = 2000);   //constructor with default parameters
4        void add_year (int);
5        int getDay() const { return day; }
6        int getMonth() const { return month; }
7        int getYear() const { return year; }
8    private:
9        int day;
10       int month;
```

```
11      int year;
12   };
```

In Lines 5 through 7, the member functions are specified as const in their prototypes. This means that the compiler does not allow these functions to modify data members of an object.

```
//-----------------------------------------------------------------------
//File: Date.cpp (Implementation file)
//This program implements the details of a Date class and test it.
//-----------------------------------------------------------------------
1   #include <iostream>
2   using namespace std;
3   #include "Date.h"
4
5   Date::Date(int d, int m, int y)
6   {  day = d; month = m; year = y;  }
7   void Date::add_year(int n)
8   {  year += n; }
9
10  int Date::getDay() const
11  {  return day;  }
12  int Date::getMonth() const
13  {  return month;  }
14  int Date::getYear() const
15  {  return year;  }
16
17  int main()
18  {
19      Date today;
20      today.add_year (18);
21      cout << today.getDay() << " / " << today.getMonth() << " / " << today.getYear() << endl;
22      return 0;
23  }
```

Result:

```
1   1 / 1 / 2018
```

The **const** keyword is placed after the (empty) parameter list in the function declarations. It indicates that these functions cannot modify the data content of a *Date* class.
const 关键字放置在函数声明中的（空）参数列表之后，它表示这些函数不能修改 Date 类的数据内容。

Naturally, the compiler will catch accidental attempts to violate this promise. If the statement in Line 7 is as follows,

```
    int getYear() const
    {
        return year++;   //error: attempt to change member value in const function
    }
```

the compiler will tell you an error information.

When a ***const*** member function is defined outside its class, the ***const*** suffix is required. In other words, the const is part of the type of *Date::getDay()*.

```
    int Date::getYear()   const
    {
        return year;   //ok
    }
```

Some objects need to be modifiable and some not. The ***const*** keyword can be used to specify that an object is not modifiable and that any attempt to modify the object should result in a compilation error. A const member function can be invoked for both const and non-const objects, whereas a non-const member function can be invoked only for non-const objects. For example,

```
    int main()
    {
        Date d1;                   //d1 is a non-const object
        const Date& d2 = d1;       //d2 is a const object
        int i = d1.getYear();      //ok
        d1.add_year(2);            //ok
        int j = d2.getYear();      //ok
        d2.add_year(2);            //error: cannot change value of const d2
        return 0;
    }
```

4.2 this Pointers

An object's member function can manipulate the object's data. How do member functions know which data members of the object to manipulate? Now, let us take a look at the following program that tests the *Date* class defined in Example 4-1.

```
1 int main()
2 {
3     Date today;
4     cout << "The size of object today:" << sizeof(today) << " Its address " << &today << endl;
5     cout << today.getDay() << " / " << today.getMonth() << " / " << today.getYear() << endl;
6     Date tomorrow(5, 4);
7     cout << "The size of object tomorrow:" << sizeof(tomorrow) << " Its address "
8         << &tomorrow << endl;
```

```
9        cout << tomorrow.getDay() << "/" << tomorrow.getMonth() << "/"
10           << tomorrow.getYear() << endl;
11       return 0;
12 }
```

Result:

```
1   The size of object today:12 Its address 0012FF74
2   1 / 1 / 2000
3   The size of object tomorrow:12 Its address 0012FF68
4   5/4/2000
```

Any object has access to its own address through a pointer called *this*. But the *this* pointer of an object is not part of the objects itself—i.e. the size of the memory taken by the *this* pointer does not affect the result of a *sizeof* operation on the objects. For example, both the sizes of the *today* and *tomorrow* objects are 12 in this example.

When a non-static member function is called for an object, the *this* pointer is passed as an implicit argument to each of the non-static member functions of a *class*, *struct*, or *union* type. However, static member functions do not have a *this* pointer. For example, the following function call

```
today.add_year(15);
```

can be interpreted this way:

```
add_year(&today, 15);
```

The object's address is available within the member function as the *this* pointer. Most usages of *this* are implicit. It is legal, though unnecessary, to explicitly use *this* when referring to members of the class. For example,

```
void Date::add_year (int n)
{
    year += n;              //ok
    this->year += n;        //ok
    (*this).year += n;      //ok
}
```

These three statements are equivalent. The expression **this* is commonly used to return the current object. For example,

```
1   #include <iostream>
2   using namespace std;
3   class Date{
4   public:
5       Date();
6       Date& setDate(int, int, int);
7       int getYear() const;
```

```
8    private:
9      int day, month, year;
10 };
11 Date::Date()
12 { day = 1;   month = 1;   year = 2015;   }
13 Date& Date::setDate(int d, int m, int y)
14 {
15    day = d;   month = m;   year = y;
16    cout << "The this address: " <<this << endl;
17    return *this;
18 }
19 int Date::getYear() const
20 {   return this->year;   }
21 int main()
22 {
23    Date today;
24    cout << "The current object address: " << &today << endl;
25    cout << today.setDate(5, 5, 2018).getYear() << endl;
26    return 0;
27 }
```

Result:

```
1   The current object address: 0045FD78
2   The this address: 0045FD78
3   2018
```

You may notice that the *setDate* function that returns a reference to a *Date* object. Line 17 returns the current object from member function *setDate* by using ***this**. Using the ***this*** pointer enables cascaded function calls in which multiple functions are invoked. In Line 25, the *today* object first invokes the *setDate* member function. Then, the *today* object with new data invokes the *getYear* function.

The **this** pointer is not an ordinary variable. Because the **this** pointer is non-modifiable, assignments to **this** are not allowed.

this 指针不是普通的变量。因为 **this** 指针是不可修改的，所以不允许给 **this** 指针赋值。

4.3 Static Members

Inside a class definition, the class members can be declared using the storage class specifier ***static*** in the class member list. Static members of a class are not associated with the objects of the class: they are independent variables with static storage duration.

> A **static member** is a variable that is part of a class, yet is not part of an object of that class. Only one copy of the static member is shared by all objects of a class in a program.
>
> 静态成员是一个类的部分的变量，但不是该类对象的一部分。程序中一个类的所有对象只共享静态成员的一个副本。

A typical use of static members is for recording data common to all objects of a class. For example, you can use a static data member as a counter to store the number of objects of a particular class type that are created. Each time a new object is created, this static data member can be incremented to keep track of the total number of objects.

Example 4-2_1: Definition of a *student* class without static members.

```
//------------------------------------------------------------------------------
//File: student.h (specification file)
//This program describes the definition of a student class.
//------------------------------------------------------------------------------
1    class student{
2    public:
3        student();
4        void print() const;
5    private:
6        int count;              //a counter recording the number of objects
7        int studentNo;          //an identifier of a student
8    };
//------------------------------------------------------------------------------
//File: student.cpp (implementation file)
//The program presents the implementation detail of the student class
//------------------------------------------------------------------------------
1    #include "student.h"
2    #include <iostream>
3    using namespace std;
4
5    student::student()
6    {
7        count = 0;
8        studentNo = count;
9    }
10
11   void student::print() const
12   {   cout << "student = " << studentNo << " count = " << count << endl;   }
//------------------------------------------------------------------------------
//File: example4-2_1.cpp
//This program creates student objects and obtains the number of student objects.
//------------------------------------------------------------------------------
1    #include "student.h"
2    #include <iostream>
```

```
3      using namespace std;
4      int main()
5      {
6          student st1;
7          st1.print();
8          cout << "***********\n";
9          student st2;
10         st1.print();
11         st2.print();
12         cout << "***********\n";
13         student st3;
14         st1.print();
15         st2.print();
16         st3.print();
17         cout << "The object size: " << sizeof(st1) << "; " << sizeof(st2)
18              << "; "<< sizeof(st3) << endl;
19         return 0;
20     }
```

Result:

```
1    student=0 count=0
2    ***********
3    student=0 count=0
4    student=0 count=0
5    ***********
6    student=0 count=0
7    student=0 count=0
8    student=0 count=0
9    The object size: 8; 8; 8
```

When we instantiate a class object, each object gets its own copy of all normal data members. In this case, since we have declared three *student* class objects, we end up with three copies of *count*—one inside *st1*, the second inside *st2*, and the third inside *st3*. They should have different values because the *count* variable is used to keep track of the total number of objects. However, they get the same value 0.

In Line 17 of example4-2_1.cpp, the size of three objects is evaluated. Their sizes are all 8 bytes because there are two *int* data members in each object.

4.3.1 Static Data Members

To keep track of the number of objects, one way to think about it is that all objects of a class share the static variables.

Example 4-2_2: Definition of a *student* class with static members.

```
//-----------------------------------------------------------------
//File: student.h (specification file)
```

//This program describes the definition of a *student* class.
//---
```
1    class student{
2    public:
3        student();
4        void print() const;
5        static int getCount();          //a static member function
6    private:
7        static int count;                //a static data member
8        int studentNo;                   //an identifier of a student
9    };
```
//---
//File: **student.cpp** (implementation file)
//The program presents the implementation detail of the *student* class.
//---
```
1    #include "student.h"
2    #include <iostream>
3    using namespace std;
4
5    int student::count = 0;              //definition outside class declaration
6
7    student::student()
8    {
9        count ++;                        //instead of count=0
10       studentNo = count;
11   }
13
14   void student::print() const
15   {
16       cout << "student = " << studentNo << " count = " << count << endl;
17   }
18
19   int student::getCount()
20   {
21       return count;
22   }
```
//---
//File: **example4-2_2.cpp**
//This program creates student objects and obtains the number of *student* objects.
//---
```
1    #include "student.h"
2    #include <iostream>
3    using namespace std;
4
5    int main()
6    {
7        student st1;
```

```
8       st1.print();
9       cout << "**********\n";
10      student st2;
11      st1.print();
12      st2.print();
13      cout << "**********\n";
14      student st3;
15      st1.print();
16      st2.print();
17      st3.print();
18      cout << "**********\n";
19      cout << "Student's number is "<< student::getCount() << endl;
20      cout << "The object size: " << sizeof(st1) << "; " << sizeof(st2)
21           << "; " <<sizeof(st3) << endl;
22      return 0;
23  }
```

Result:

```
1   student=1 count=1
2   **********
3   student=1 count=2
4   student=2 count=2
5   **********
6   student=1 count=3
7   student=2 count=3
8   student=3 count=3
9   **********
10  Student's number is 3
11  The object size: 4; 4; 4
```

Since the *count* variable is a static data member, *count* is shared among all objects of the class. Consequently, *st1.count* is the same as *st2.count* after object *st2* is declared in Line 10 of the example4-2_2.cpp file.

Although you can access static members through objects of the class type, this is somewhat misleading. *st1.count* implies that *count* belongs to *st1*, which is not really the case. *count* does not belong to any object. In fact, *count* exists even if there are no instantiated objects of the class!

Observing, the result from Line 11 of example4-2_2.cpp, although *count* is a data member of the *student* class, it does not take a space of the object in memory. Therefore, each object's size is only 4 bytes.

Initializing Static Data Members

Since static data members are not part of the individual objects, you must explicitly define the static member if you want to initialize it to a non-zero value. The ***static*** keyword is only used with the declaration of a static member, inside the class definition, but not with the definition of that static member.

The following statement in Example 4-2_2 initializes the static member to 0:

```
int student::count = 0;
```

This initializer should be placed in the implementation file for the class (e.g. student.cpp of Example 4-2_2).

The declaration of a static data member in the member list of a class is not a definition. You must define the static members outside of the class declaration, in namespace scope, even if the static members are private.
类成员列表中的静态数据成员声明不是一个定义，你必须在命名空间范围内类声明之外定义的静态成员，即使静态成员是私有的。

4.3.2 Static Member Functions

While we create a normal public member function to access the *count* variable in Example 4-2_2, we need to instantiate an object of the class to use the function. Like static data members, static member functions are not attached to any particular object. Here is the example above with a static member function:

```
static int getCount()
{   return count;   }
```

Static member functions can be called directly by using the class name and the scope operator because they are not attached to any particular object. Like static data members, they can also be called through objects of the class type, though this is not recommended. In Example 4-2_2, the static member function is invoked as follows:

```
1    int main() {
2        student st1;
3        st1.print();
4        cout << "***********\n";
5        student st2;
6        st1.print();
7        st2.print();
8        cout << "***********\n";
9        student st3;
10       st1.print();
11       st2.print();
12       st3.print();
13       cout << "***********\n";
14       cout << "Student's number is " << student::getCount() << endl;
```

```
15        return 0;
16    }
```

Static member functions have two interesting features worth noting. First, since static member functions are not attached to any object, they do not have the ***this*** pointer! This makes sense when you think about it—the ***this*** pointer always points to the object that the member function is working on. Static member functions do not work on an object, so the ***this*** pointer is not needed.

Second, static member functions can only access static member variables. They cannot access non-static data members. This is because non-static data member must belong to a class object while the static member functions have no class object to work with!

Static member functions
- have no **this** pointer because they are not associated with any objects.
- cannot access non-static data member.
- cannot be virtual and const.

静态成员函数
- 因为它们不与任何对象相关联，故没有 this 指针。
- 不能存取非静态数据成员。
- 不能定义为虚函数和常量函数。

Think These Over

1. Why do static data members of a class differ from the ordinary data members?
2. How do we obtain static data members?

4.4 Free Store

C++ has several distinct memory areas where objects and non-object values may be stored, and each area has different characteristics shown in Table 4-1.

Table 4-1 Characteristics of different memory areas

Memory Area	Characteristics and Object Lifetime
Const data	The const data area stores string literals and other data whose values are known at compile time. No objects of a class type can exist in this area. All data in this area is available during the entire lifetime of the program. Further, all of this data is read-only, and the results of trying to modify it are undefined.
Stack	The stack stores automatic variables. Typically, the allocation is much faster than for dynamic storage (heap or free store) because a memory allocation involves only pointer increment rather than more complex management. Objects are constructed immediately after the memory is allocated and destroyed immediately before the memory is deallocated, so there is no opportunity for programmers to directly manipulate allocated but uninitialized stack space.

(continued)

Memory Area	Characteristics and Object Lifetime
Free store	The free store is one of the two dynamic memory areas, allocated/freed by new/delete. Object lifetime can be less than the time the storage is allocated; that is, free store objects can have memory allocated without being immediately initialized, and can be destroyed without the memory being immediately deallocated. During the period when the storage is allocated but outside the object's lifetime, the storage may be accessed and manipulated through a void* but none of the proto-object non-static members or member functions may be accessed, have their addresses taken, or be otherwise manipulated.
Heap	The heap is the other dynamic memory area, allocated/freed by malloc/free and their variants. Note that while the default global new and delete might be implemented in terms of malloc and free by a particular compiler, the heap is not the same as free store and memory allocated in one area cannot be safely deallocated in the other. Memory allocated from the heap can be used for objects of class type by placement-new construction and explicit destruction. If so used, the notes about free store object lifetime apply similarly here.
Global / Static	Global or static variables and objects have their storage allocated at program startup, but may not be initialized until after the program has begun executing. For instance, a static variable in a function is initialized only the first time program execution passes through its definition. The order of initialization of global variables across translation units is not defined, and special care is needed to manage dependencies between global objects (including class statics). As always, uninitialized proto-objects' storage may be accessed and manipulated through a void* but no non-static members or member functions may be used or referenced outside the object's actual lifetime.

The free store is a pool of memory available for you to allocate (and deallocate) storage for objects during the execution of your program. The ***new*** and ***delete*** operators are used to allocate and deallocate free store, respectively.

The return value from ***new*** is a memory address. It must be assigned to a pointer. To create an integer in the free store, you might write

```
int *ptr = new int;
int *ptr1 = new int(8);
char *p = new char[10];
```

You can, of course, initialize the pointer at its creation with

```
int *ptr;
ptr = new int;
```

In either case, the *ptr* variable now points to an integer in the free store. You can use this like any other pointer to a variable and assign a value into that area of memory by writing

```
*ptr =20;
```

This means, "put 20 at the value in *ptr*," or "assign the value 20 to the area in the free store to where *ptr* points".

If using the ***new*** keyword cannot create a memory on the free store (memory is, after all, a limited resource), it returns the null pointer. You must check your pointer for null each time you request new memory.

Each time you allocate memory using the **new** keyword, you must check to make sure the pointer is not null.
每次使用 **new** 关键字分配内存时，必须检查以确保指针不是空的。

When you are done with your area of memory, you must call the ***delete*** statement on the pointer. The ***delete*** statement returns the memory to the free store. Remember that the pointer itself—as opposed to the memory to which it points—is a local variable. When the function in which it is declared returns, that pointer goes out of scope and is lost. The memory allocated by using ***new*** is not freed automatically, however. That memory becomes unavailable—a situation called a ***memory leak***. The name is given to describe that the part of memory cannot be recovered until the program ends. It is as though the memory has leaked out of your computer.

To restore the memory to the free store, you use the ***delete*** keyword. For example,

```
delete ptr;
delete ptr1;
delete []p;
```

When you delete the pointer, what you are really doing is freeing up the memory whose address is stored in the pointer.

Objects by Using *new* and *delete* on Free Store

Just as you can create a pointer to an integer, you can create a pointer to any object. If you have declared an object of class *Date*, you can declare a pointer to that class and instantiate a *Date* object on the free store, just as you can make one on the stack.

Example 4-3: The *Date* objects by using *new* and *delete*.

```
//---------------------------------------------------------------
//File: Date.h
//This program describes the definition of a Date class.
//---------------------------------------------------------------
1 class Date {
2 public:
3     Date();                    //default constructor
4     Date(int, int, int);       //constructor
5     ~Date();                   //destructor
6     int getDay() const;
7     int getMonth() const;
8     int getYear() const;
9 private:
10    int day, month, year;
11 };
//---------------------------------------------------------------
```

· 99 ·

```
//File: Date.cpp
//The program presents the implementation detail of the Date class.
//---------------------------------------------------------------------------------------
1    #include "Date.h"
2    #include <iostream>
3    using namespace std;
4
5    Date::Date()
6    {
7        day = 1; month = 3; year = 2015;
8        cout << "Default constructor...\n";
9    }
10
11   Date::Date(int d, int m, int y)
12   {
13       day = d; month = m; year = y;
14       cout << "constructor...\n";
15   }
16
17   Date::~Date()
18   {   cout << "Destructor...\n";   }
19
20   int Date::getDay() const
21   {   return day;   }
22   int Date::getMonth() const
23   {   return month;   }
24   int Date::getYear() const
25   {   return year;   }
```
//---
//File: example4_3.cpp
//This program creates Date objects by using new on the free store and accesses their members.
//---
```
1    #include "Date.h"
2
3    int main()
4    {
5        Date day;
6        Date today(18, 5, 2015);
7        Date *dPtr;
8        dPtr = &day;
9
10       cout << "Constructing objects by new....\n";
11       Date *dPtr1 = new Date;
12       Date *dPtr2 = new Date(20, 3, 2000);
13
14       cout << dPtr->getDay() << "/" << dPtr->getMonth() << "/" << dPtr->getYear() << endl;
15       cout << dPtr1->getDay() << "/" << dPtr1->getMonth() << "/" << dPtr1->getYear() << endl;
```

```
16        cout << dPtr2->getDay() << "/" << dPtr2->getMonth() << "/" << dPtr2->getYear() << endl;
17
18        cout << "Destroying objects by delete....\n";
19        delete dPtr1;
20        delete dPtr2;
21
22        cout << "Ending execution...\n";
23        return 0;
24   }
```

Result:

1	Default constructor...
2	constructor...
3	Constructing objects by new....
4	Default constructor...
5	constructor...
6	1/3/2015
7	1/3/2015
8	20/3/2000
9	Destroying objects by delete....
10	Destructor...
11	Destructor...
12	Ending execution...
13	Destructor...
14	Destructor...

In Line 5 in example4_3.cpp, the *day* object is created on the stack while the default constructor is called. In Line 6, the *today* object is created on the stack whilst the constructor is called. In Line 11, the *dPtr*1 object is created in the free store. The default constructor sets its *day*, *month* and *year* to 1, 3, 2015. In Line 12, the *dPtr*2 object is created in the free store. The constructor sets its *day*, *month* and *year* to 20, 3, 2000. When *dPtr*1 and *dPtr*2 are deleted in Lines 19 and 20, their destructors are called. The destructor deletes each of its member pointers. Since both *dPtr*1 and *dPtr*2 are pointers, the arrow operator (->) is used to access the member function *print* in Lines 14, 15 and 16.

4.5 Object Members

4.5.1 Definition of Object Members

Sometimes, a class has a data member which is not a simple built-in data type but an aggregate built-in data. Such data member is defined as another class object. When a class has objects of other classes as members, such capability is called ***composition***. Composition is a special case of the *aggregation* relationship.

> An **object member** of a class is a data member that is defined as another class object. The object member has two forms in the class, that is, **composition** and **aggregation.**
>
> Composition and aggregation are forms of software reusability, a "**has a**" relationship, and represent an ownership relationship between two objects.
>
> Aggregation differs from composition in that it does not imply ownership.
>
> 类对象成员是将另一个类的对象定义为该类的数据成员。对象成员在类中有两种形式，即**组合**和**聚合**。
>
> 组合和聚合是软件可重用性的一种形式，是一种"**has a**"关系，并且表示两个对象之间的拥有关系。
>
> 聚合不同于组合，因为它并不意味着具有拥有关系。

When an object is created, its constructor is called automatically, so we need to specify how arguments are passed to member-object constructors. Member objects are constructed in the order in which they are declared (not in the order they are listed in the constructor's member initializing list) and before their enclosing class objects are constructed.

For example, we define a *Student* class that has the properties of age, birthday and body status. The birthday is defined as the *Date* class type. The body status is defined as the *Health* class type that includes a student's height and weight. The birthday and body status are user-defined types. This means that the *Student* class owns two other class objects. "A student has a birthday" is an *aggregation* relationship between the *Student* class and the *Date* class because a birthday may be shared by other students, whereas "a student has a body status" is a *composition* relationship between the *Student* class and the *Health* class. The UML diagram of their class relationship is shown in Figure 4-1. In UML, a filled diamond denotes the composition relationship between the *Student* class and the *Health* class, and an empty diamond denotes the aggregation relationship between the *Student* class and the *Date* class.

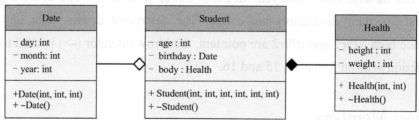

Figure 4-1   The UML diagram of the class ownership

Example 4-4: Definition of member objects.

```
//-------------------------------------------------------------------------
//File: student.h
//This program describes the definition of a Student class with Date and Health objects.
//-------------------------------------------------------------------------
```

```
1    class Date {
2      public:
3          Date(int, int, int);
4          ~Date();
5      private:
6          int day, month, year;
7    };
8
9    class Health {
10     public:
11         Health(int, int);
12         ~Health();
13     private:
14         int height;
15         int weight;
16   };
17
18   class Student {
19     public:
20         Student(int, int, int, int, int, int);
21         ~Student();
22     private:
23         int age;
24         Date birthday;              //object of class Date
25         Health body;                //object of class Health
26   };
```

//---
//File: **student.cpp**
//The program presents the implementation detail of three classes *Date*, *Health* and *Student*.
//---

```
1    #include "student.h"
2    #include <iostream>
3    using namespace std;
4
5    Date::Date(int d, int m, int y)
6    {
7        day = d; month = m; year = y;
8        cout << "Date class is created" << endl;
9    }
10
11   Date::~Date()
12   {
13       cout << "Date class is destroyed" << endl;
14   }
15
16   Health::Health(int h, int w)
17   {
```

```
18          height = h; weight = w;
19          cout << "Health class is created" << endl;
20      }
21
22      Health::~Health()
23      {
24          cout << "Health class is destroyed" << endl;
25      }
26
27      Student::Student(int a, int d, int m, int y, int h, int w) :
28                  birthday(d, m, y), body(h, w)    <--------initialization list
29      {
30          age = a;
31          cout << "Student class is created" << endl;
32      }
33
34      Student::~Student()
35      {
36          cout << "Student class is destroyed" << endl;
37      }
```
//---
//File: example4_4.cpp
//This program demonstrates how two member objects are initialized.
//---
```
1       #include "student.h"
2       int main()
3       {
4           Student st(20, 11, 2, 1990, 180, 75);
5           return 0;
6       }
```

Result:

1	Date class is created
2	Health class is created
3	Student class is created
4	Student class is destroyed
5	Health class is destroyed
6	Date class is destroyed

The program above defines three classes *Date*, *Health* and *Student* to demonstrate objects as members of other class. Class *Student* includes private data members *age*, *birthday* and *body*. Members *birthday* and *body* are the objects of class *Date* and *Health*. The program instantiates a *Student* object and initializes it.

The *Student* constructor is defined as follows:

Student::Student(int a, int d, int m, int y, int h, int w) :
                birthday(d, m, y), body(h, w)
    { age = a; }   → Object of the *Health* class
                                           → Object of the *Date* class

The constructor takes six parameters. The colon (:) in the header separates the member initialization list (also called the member initializer) from the parameter list. The member initialization list specifies the *Student* parameters being passed to the constructors of the objects. Arguments *d*, *m* and *y* are passed to the *birthday* constructor, and arguments *h* and *w* are passed to the *body* constructor.

4.5.2 The Order of Constructors and Destructors for Member Objects

For creating objects, the members' constructors are invoked before the body of the containing class' own constructor is executed. The constructors are invoked in the order in which the members are declared in the class rather than the order in which the members appear in the initialization list.

For destroying objects, the body of that object's own destructor is executed first and then the members' destructors are executed in reverse order of the declaration.

4.5.3 Object Members with Default Constructors

If a member constructor needs no arguments, the member does not need to be mentioned in the member initializer. For example,

```
//----------------------------------------------------------------
//File: example4_4_1.cpp
//This program demonstrates the initialization of member objects with a default constructor.
//----------------------------------------------------------------
1    #include <string>
2    using namespace std;
3
4    class Score {
5    public:
6        Score():grade(0){}
7    private:
8        int grade;
9    };
10
11   class Student {
12   public:
13       Student(char* nstr);
14       ~Student();
15   private:
```

```
16        char* name;
17        Score sc;                //object of class Score
18    };
19
20    Student::Student(char* nstr): sc()
21    {
22        name = new char[strlen(nstr) + 1];
23        strcpy(name, nstr);
24    }
25    Student::~Student()
26    {
27        Delete name;
28    }
29    int main()
30    {
31        Student st("wang");
32        return 0;
33    }
```

To ensure fulfilling the initialization requirements for the member objects, one of the following conditions must be met:
- The contained object's class requires no constructor.
- The contained object's class has an accessible default constructor.
- The containing class' constructors all explicitly initialize the contained object.

为了确保成员对象的初始化要求，必须满足下列条件之一：
- 包含对象的类不需要构造函数。
- 包含对象的类有一个可访问的默认构造函数。
- 包含类的构造函数都显式地初始化包含的对象。

4.5.4 Class Members by Using Initializers

Member initializers are essential for types for which initialization differs from the assignment, that is, for **member objects** of classes without a default constructor, for **const** members, and for **reference** members.

For example,

```
//-----------------------------------------------------------------------
//File: example4_4_2.cpp
//This program demonstrates the initialization of special members.
//-----------------------------------------------------------------------
1     class A {
```

```
 2    public :
 3      A(int dd, int mm, int yy, int& a) : i(10), d(dd, mm, yy), pc(a){}
 4    private:
 5      const int i;      //const member
 6      Date d;           //member object
 7      int& pc;          //reference member
 8    };
 9
10    int main()
11    {
12      int i;
13      A a(23, 10, 2000, i);
14      return 0;
15    }
```

Above the constructor: initialization → reference variable → object of class Date → const variable

Think These Over

1. How is an object member of the class initialized?

2. What is the purpose of composition?

3. How to draw the UML class diagram of example 4_4_1.cpp?

4.6 Copy Members

When declaring a class object, you can initialize it by using the values of an existing object of the same class.

For example, consider the following statement:

```
Date today (18, 11, 2010);
Date day = today;   or   Date day(today);
```

Object *today* has been declared. The new object *day* is declared and is also initialized by using the value of object *today*, that is, the values of the data members of *today* are copied into the corresponding data members of *day*, as shown in Figure 4-2.

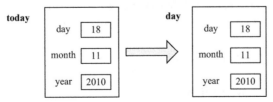

Figure 4-2 Objects *today* and *day*

When the class object is initialized by default member-wise initialization, this default initialization is due to the constructor, called the **copy constructor**.

A **copy constructor** is a special constructor for creating a new object as a copy of an existing object.

> 当类对象的初始化以默认逐一成员进行初始化时，这种默认初始化是由构造函数（称为**拷贝构造函数**）造成的。
>
> **拷贝构造函数**是一个特殊构造函数，它通过已有的对象以拷贝方式创建新对象。

4.6.1 Definition of Copy Constructors

The copy constructor takes a ***reference*** to a ***const*** parameter. It is a *const* to guarantee that the copy constructor does not change it. In addition, it is a *reference* because a value parameter would require making a copy, which would invoke the copy constructor, therefore making a copy of its parameter.

Example 4-5: Definition of a copy constructor.

```
//-----------------------------------------------------------------------
//File: Date.h
//This program describes the definition of a Date class.
//-----------------------------------------------------------------------
16    class Date {
17    public:
18        Date(int = 1, int = 1, int = 2000);
19        Date(const Date&);                    //copy constructor
20        ~Date();
21        void print() const;
22    private:
23        int day, month, year;
24    };
//-----------------------------------------------------------------------
//File: Date.cpp
//The program presents the implementation detail of the Date class.
//-----------------------------------------------------------------------
1     #include "Date.h"
2     #include <iostream>
3     using namespace std;
4
5     //Definition of constructor with default arguments
6     Date::Date(int d, int m, int y) : day(d), month(m), year(y)
7     {
8         cout << "constructing an object...\n";
9     }
10    //Definition of copy constructor
11    Date::Date(const Date& date)
12    {
13        day = date.day;
14        month = date.month;
15        year = date.year;
16        cout << "constructing a copy object...\n";
```

```
17  }
18  //Definition of destructor
19  Date::~Date()
20  {
21      cout << "destroying an object...\n";
22  }
23  void Date::print() const
24  {
25      cout << day<< "/" << month << "/" << year << endl;
26  }
```
//--
//File: **example4_5.cpp**
//This program demonstrates how the copy constructor works.
//--
```
1   #include "Date.h"
2   #include <iostream>
3   using namespace std;
4   Date f(Date d)
5   {
6       cout << "This is a sub-function...\n";
7       d.print();
8       return d;
9   }
10
11  int main()
12  {
13      Date day1(7, 8, 2010);
14      Date day2 = day1;          //declaration of copy object
15      Date day3(day1);           //declaration of copy object
16      day1.print();
17      day2.print();
18
19      cout << endl;
20      day3 = f(day2);            //copy constructor initializes formal parameter
21      cout << endl;
22      cout << "Ending the main function...\n";
23      return 0;
24  }
```

Result:

```
1   constructing an object...
2   constructing a copy object...
3   constructing a copy object...
4   7/8/2010
5   7/8/2010
6
7   constructing a copy object...
```

```
8   This is a sub-function...
9   7/8/2010
10  constructing a copy object...
11  destroying an object...
12  destroying an object...
13
14  Ending the main function...
15  destroying an object...
16  destroying an object...
17  destroying an object...
```

The above is an example of a copy constructor for the *Date* class, which does not really need one because the default copy constructor's action of copying data would work fine. Still, it shows how it works.

When Line 14 in example4_5.cpp declares a copy object *day2*, the copy constructor is called. From the result of Example 4-5, you can find out that the copy constructor is called when the *f* function is called in Line 20 and when the *d* object is returned in the *f* function.

Therefore, the copy constructor happens in the following cases:

(1) Declaring an object which is initialized by another object with the same type, for example,

```
Date day1(7, 8, 2018);
Date day2 = day1;
```

(2) Initializing a value parameter by its corresponding argument (copy an object to pass it as an argument to a function), for example,

```
f(day2);
```

(3) Returning an object in a function (copy an object to return it from a function), for example,

```
Date f()
{
    Date d(7, 8, 2018)
    return d;
}
```

C++ calls a copy constructor to make a copy of an object for each case above. If there is no copy constructor defined for the class, C++ uses the default copy constructor which copies every field, that is, makes a ***shallow copy***.

4.6.2 Shallow Copy and Deep Copy

Shallow Copy

A ***shallow copy*** means that C++ copies each member of the class individually using the assignment operator (=). When classes are simple (e.g. do not contain any dynamically allocated memory), this works very well like the example above.

If the object has no pointers to dynamically allocated memory, a shallow copy is probably sufficient. In that case, the default copy constructor, the default assignment operator, and the default destructor will work just fine, meaning that you do not need to write your own.

Now let's make some changes for Example 4-5—remove the definition of *copy constructor* from Example 4-5. The program still works well. The changed program is illustrated in Example 4-6.

Example 4-6: Shallow copying the class members by using a default copy constructor.

```
    //----------------------------------------------------------------
    //File: Date.h
    //This program describes the definition of a Date class.
    //----------------------------------------------------------------
2   class Date {
3   public:
4       Date(int = 1, int = 1, int = 2000);
5       ~Date();
6       void print() const;
7   private:
8       int day, month, year;
9   };
    //----------------------------------------------------------------
    //File: Date.cpp
    //The program presents the implementation detail of the Date class.
    //----------------------------------------------------------------
2   #include "Date.h"
3   #include <iostream>
4   using namespace std;
5
6   //Definition of constructor with default arguments
7   Date::Date(int d, int m, int y):day(d), month(m), year(y)
8   {
9       cout << "constructing an object...\n";
10  }
11  //Definition of destructor
12  Date::~Date()
13  {
14      cout << "destroying an object...\n";
15  }
16  void Date::print() const
17  {
18      cout << day << "/" << month << "/" << year << endl;
19  }
    //----------------------------------------------------------------
```

```
//File: example4_6.cpp
//This program demonstrates how the default copy constructor works.
//-------------------------------------------------------------------
1   #include "Date.h"
2   #include <iostream>
3   using namespace std;
4
5   Date f(Date d)
6   {
7       cout << "This is a sub-function...\n";
8       d.print();
9       return d;
10  }
11  int main()
12  {
13      Date day1(7, 8, 2010);
14      Date day2 = day1;          //calling a default copy constructor
15      Date day3(day1);           //calling a default copy constructor
16      day1.print();
17      day2.print();
18
19      cout << endl;
20      day3 = f(day2);            //copy constructor initializes formal value parameter
21      cout << endl;
22      cout << "Ending the main function...\n";
23      return 0;
24  }
```

Result:

```
1   constructing an object...
2   7/8/2010
3   7/8/2010
4
5   This is a sub-function...
6   7/8/2010
7   destroying an object...
8   destroying an object...
9
10  Ending the main function...
11  destroying an object...
12  destroying an object...
13  destroying an object...
```

Although the result above is different from that of Example 4-5, the program works well. The reason is that the default copy constructor provided by the system is used when the copy objects are declared, e.g. in Lines 14 and 15 of file example4_6.cpp, is passed as a parameter in Line 5 or returning value in Line 9.

However, when designing classes that handle dynamically allocated memory, member-wise (shallow) copying can get us in a lot of trouble! This is because the standard pointer assignment operator just copies the address of the pointer—it does not allocate any memory or copy the contents being pointed to! Let's take a look at the following example.

Example 4-7: Shallow copying a pointer to a data member in a class.

//---
//File: **Date.h**
//This program describes the definition of a *Date* class.
//---

```
1    class Date {
2    public:
3        Date(int, int, int, char*);
4        ~Date();
5        void print() const;
6    private:
7        int day, month, year;
8        char *name;                      //char pointer
9    };
```

//---
//File: **Date.cpp**
//The program presents the implementation detail of the *Date* class.
//---

```
1    #include "Date.h"
2    #include <iostream>
3    #include <string.h>
4    using namespace std;
5    //Definition of constructor
6    Date::Date(int d, int m, int y, char* nstr):day(d), month(m), year(y)
7    {
8        name = new char[strlen(nstr) + 1];   //allocate memory dynamically on free store
9        strcpy(name, nstr);
10
11       cout << name << " is constructing...\n";
12   }
13   //Definition of destructor
14   Date::~Date()
15   {
16       cout << name << " is destroying...\n";
17       delete name;                          //deallocating memory
18   }
19   void Date::print() const
20   {
21       cout << day << "/" << month << "/" << year << endl;
22   }
```

• 113 •

```
//---------------------------------------------------------------
//File: example4_7.cpp
//This program demonstrates what problems shallow copying causes.
//---------------------------------------------------------------
1    #include "Date.h"
2    #include <iostream>
3    using namespace std;
4
5    void f(Date d)
6    {
7        cout << "This is a sub-function...\n";
8        d.print();
9    }
10   int main()
11   {
12       Date day1(7, 8, 2010, "day1");
13       Date day2 = day1;                    //declaration of copy object
14       Date day3(12, 5, 2000, "day3");
15
16       day1.print();
17       day2.print();
18
19       cout << endl;
20       f(day2);
21       cout << endl;
22       cout << "Ending the main function...\n";
23
24       return 0;
25   }
```

The program above seems to have no problem, but it contains an insidious problem that will cause the program to crash! When it is executed, an error message is prompted.

Let's break down this example. Line 13 is the same as that in Example 4-6 and seems harmless enough as well, but it's actually the source of our problem causing the error message. Why? As we mentioned above, object *day2* is declared by using existing object *day1*. This copy is executed by using the default copy constructor (because we haven't defined our own), which does a shallow pointer copy for data member *day1.name*. Since a shallow pointer copy just copies the address of the pointer, the address of *day1.name* is copied to *day2.name*. As a result, now both members are pointing to the same piece of memory, as shown in Figure 4-3! When reaching the end of the example4_7.cpp program, the *day2* object is destroyed and the dynamically allocated memory, which *day2.name* points to, is cleaned up. When the *day1* object is destroyed, the *day1.name* points to the deleted (invalid) memory! Thus, you see the error message. Likewise, the same error occurs when calling the *f* sub-function.

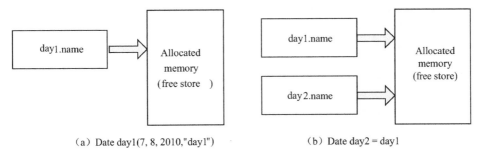

(a) Date day1(7, 8, 2010,"day1")   (b) Date day2 = day1

Figure 4-3   Shallow copy for *day2=day1*

Deep Copy

The root of this problem is the shallow copy done by the copy constructor—doing a shallow copy on pointer values in a copy constructor or an overloaded assignment operator (=) usually means trouble. The solution to this problem is to do a deep copy on any non-null pointers being copied.

A ***deep copy*** duplicates the object or variable being pointed to so that the destination (the object being assigned to) receives its own local copy (see Figure 4-4). In this way, the destination can do whatever it wants to in its local copy while the object being copied from will not be affected. Doing deep copies requires that we write our own copy constructors and overloaded assignment operators (see Chapter 5).

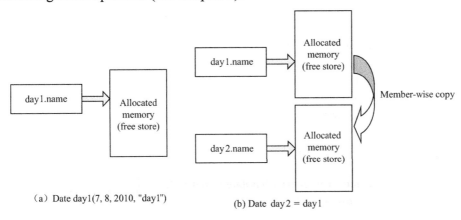

(a) Date day1(7, 8, 2010, "day1")   (b) Date day2 = day1

Figure 4-4   Deep copy for *day2 =day1*

Let's go ahead and show how this is done to solve the problem.

Example 4-8: Deep copying a pointer to a data member in a class.

```
//---------------------------------------------------------------
//File: Date.h
//This program describes the definition of a Date class.
//---------------------------------------------------------------
1    class Date {
2    public:
```

```
3          Date(int, int, int, char*);
4          Date(const Date&);          //copy constructor
5          ~Date();
6
7          void print() const;
8     private:
9          int day, month, year;
10         char *name;                 //char pointer
11    };
```
//--
//File: **Date.cpp**
//The program presents the implementation detail of the *Date* class.
//--

```
1     #include "Date.h"
2     #include <iostream>
3     using namespace std;
4
5     //Definition of constructor with default arguments
6     Date::Date(int d, int m, int y, char* nstr):day(d), month(m), year(y)
7     {
8          name = new char[strlen(nstr) + 1];
9          strcpy(name, nstr);
10
11         cout << name << " is constructing...\n";
12    }
13    //Definition of copy constructor
14    Date::Date(const Date& date)
15    {
16         day = date.day;
17         month = date.month;
18         year = date.year;
19
20         if(date.name != NULL)
21         {
22              name = new char[strlen(date.name) + 1];        ⎫
23              strcpy(name, date.name);                       ⎬  deep copying
24         }                                                   ⎪
25         else                                                ⎭
26              name = NULL;
27         cout << "constructing a copy object...\n";
28    }
29    //Definition of destructor
30    Date::~Date()
31    {
32         cout << name << " is destroying...\n";
33         delete name;
34    }
```

```
35    void Date::print() const
36    {
37        cout << day << "/" << month << "/" << year << endl;
38    }
```
//--
//File: **example4_8.cpp**
//This program demonstrates how deep copying works.
//--
```
1     #include "Date.h"
2     #include <iostream>
3     using namespace std;
4
5     void f(Date d)
6     {
7         cout << "This is a sub-function...\n";
8         d.print();
9         //return d;
10    }
11
12    int main()
13    {
14        Date day1(7, 8, 2010, "day1");
15        Date day2 = day1;           //declaration of copy object
16        Date day3(12, 5, 2000, "day3");
17
18        day1.print();
19        day2.print();
20
21        cout << endl;
22        f(day2);
23        cout << endl;
24        cout << "Ending the main function..\n";
25
26        return 0;
27    }
```

Result:

```
1     day1 is constructing...
2     constructing a copy object...
3     day3 is constructing...
4     7/8/2010
5     7/8/2010
6
7     constructing a copy object...
8     This is a sub-function...
9     7/8/2010
10    day1 is destroying...
```

```
11
12    Ending the main function...
13    day3 is destroying...
14    day1 is destroying...
15    day1 is destroying...
```

A class that requires deep copies generally needs:

• A constructor to either make an initial allocation or set the pointer to NULL.

• A destructor to delete the dynamically allocated memory.

• A copy constructor to make a copy of the dynamically allocated memory.

• An overloaded assignment operator to make a copy of the dynamically allocated memory.

需要深度拷贝的类一般需要:

- 构造函数要么将进行初始分配，要么将指针设置为 NULL。
- 析构函数用于删除动态分配的内存。
- 拷贝构造函数用来复制动态分配的内存。
- 重载赋值操作符用来复制动态分配的内存。

Think These Over

1. What situations require defining a copy constructor of a class?
2. What is the difference between shallow copy and deep copy?

4.7 Arrays of Objects

Suppose that you need to perform a number of objects. All objects are declared from the same class. A convenient way to store these objects is to use an array.

4.7.1 Initialize an Object Array by Using a Default Constructor

Suppose that you declare an array of 4 class objects, for example, 4 *Date* objects. you need to display their days, months and years. You have to declare one array *arrayDate* of 4 elements each, wherein each element is an object of type *Date*.

```
Date arrayDate[4]
```

The statement above creates the array of four objects *arrayDate[0]*, *arrayDate[1]*,..., *arrayDate[3]*.

In this case, it is impractical to specify different constructors for each element. Therefore, the default constructor is used to initialize each (array) class object. If a class has constructors and you declare an object of that class, the class should have the default constructor.

In the example above, each *arrayDate* is initialized by calling *Date::Date()*. Thus, you can use the member functions of class *Date* to display their dates for each object. For example, you can write as follows:

> arrayDate[0].print();

You can also use a loop to display the dates of four objects, such as the following:

> for (int i = 0; i < 4; i++)
>   arrayDate[i].print();

The destructor for each constructed elements of an array is invoked when that array is destroyed. If the array of class objects is constructed by using ***new***, then you must use ***delete*** explicitly to deallocate memory. Otherwise, it leads to a problem. For example,

> Date *day = new Date[4];   //allocate an array
> //....
> Delete []day;   //free an array

Example 4-9: Initializing arrays of objects by using a default constructor.

```
//---------------------------------------------------------------
//File: Date.h
//This program describes the definition of a Date class with a default constructor.
//---------------------------------------------------------------
1    class Date{
2    public:
3         Date();
4         ~Date();
5         void print() const;
6    private:
7         int day, month, year;
8         static int count;
9    };
//---------------------------------------------------------------
//File: Date.cpp
//The program presents the implementation detail of the Date class.
//---------------------------------------------------------------
1    #include "Date.h"
2    #include <iostream>
3    using namespace std;
4
5    int Date::count =0;
6    Date::Date()
7    {
8         day =1; month = 1; year=2000;
9         cout << "ArrayDate[" << count << "] is constructed\n";
10        count++;
11   }
12   Date::~Date()
```

```
13      {
14          count--;
15          cout << "ArrayDate[" << count << "] is destroyed\n";
16      }
17      void Date::print() const
18      {   cout << day << "/" << month << "/" << year << endl;   }
```
//--
//File: example4_9.cpp
//This program demonstrates how to declare object arrays and access the element of the array.
//--

```
1       #include "Date.h"
2
3       int main()
4       {
5           Date arrayDate[4];                    //array objects
6           Date *day1 = new Date;                //an object pointer
7           Date *day2 = new Date[4];             //4 object pointers
8           cout << endl;
9
10          for(int i = 0 ; i < 4; i++)
11          {
12              cout << "arrayDate[" << i << "]'s date is ";
13              arrayDate[i].print();             //use member function
14              cout << "point to array of day[" << i << "] date is ";
15              day2[i].print();
16          }
17          cout << endl;
18
19          delete day1;                          //free object day1
20          delete []day2                         //free 4 objects day2
21          return 0;
22      }
```

Result:

```
1   ArrayDate[0] is constructed
2   ArrayDate[1] is constructed
3   ArrayDate[2] is constructed
4   ArrayDate[3] is constructed
5   ArrayDate[4] is constructed
6   ArrayDate[5] is constructed
7   ArrayDate[6] is constructed
8   ArrayDate[7] is constructed
9   ArrayDate[8] is constructed
10
11  arrayDate0[0]'s date is 1/1/2000
12  point to array of day[0] date is 1/1/2000
13  arrayDate0[1]'s date is 1/1/2000
```

```
14   point to array of day[1] date is 1/1/2000
15   arrayDate0[2]'s date is 1/1/2000
16   point to array of day[2] date is 1/1/2000
17   arrayDate0[3]'s date is 1/1/2000
18   point to array of day[3] date is 1/1/2000
19
20   ArrayDate[8] is destroyed
21   ArrayDate[7] is destroyed
22   ArrayDate[6] is destroyed
23   ArrayDate[5] is destroyed
24   ArrayDate[4] is destroyed
25   ArrayDate[3] is destroyed
26   ArrayDate[2] is destroyed
27   ArrayDate[1] is destroyed
28   ArrayDate[0] is destroyed
```

The beginning of this section stated that if you declare an array of class objects while the class has the constructor(s), then the class should have a default constructor, e.g. *Date()* in the header file (date.h) in Example 4-9.

4.7.2 Initialize an Object Array by Using Constructors with Parameters

Suppose that the *Date* class has a constructor with parameters instead of a default constructor in the header file (Date.h) of Example 4-9, like the following:

```
1    //Date.h
2
3    class Date{
4    public:
5        Date(int, int, int);
6        ~Date();
7        void print() const;
8    private:
9        int day, month, year;
10       static int count;
11   };
```

then the *arrayDate* object is declared in the user program (e.g., example4_9.cpp) and initialized in the following way:

```
Date arrayDate[4] = { Date(6,4,1900),
                      Date(6,4,2000),
                      Date(15, 9, 2003),
                      Date(23, 7, 2010)};
```

In fact, the expression *Date(6, 4, 1900)* creates an anonymous object of the *Date* class; initializes its date members to 6, 4, 1900, respectively; and then uses member-wise copy to initialize the *arrayDate[0]* object.

4.8 Friends

As discussed in the previous sections on access specifiers, to access data members *day*, *month* and *year* from the outside of the class by the *main* function, the data members are declared as private inside the *Date* class in the previous examples, because the *main* function is not a class member that will not be able to access the private data. To access the private data members, data members are put into the public part of the class. However, this breaks the information hiding of the class. In fact, a programmer may have a situation where he or she would need to access private data from non-member functions. For handling such cases, the concept of *friend* functions is a useful tool.

> A **friend** of a class is a function or class that is not a member of the class, but is granted the same access to the class as the members of the class.
> 类的友元是一个函数或类，它不是类的成员，但是被赋予与类的成员相同的访问权。

Functions declared with the *friend* specifier in a class member list are called *friend functions* of that class. Classes declared with the *friend* specifier in the member list of another class are called *friend classes* of that class.

Difference between Ordinary Member Functions and Friend Functions

An ordinary member function of a class specifies three logically distinct things:

• the function can access the private part of the class,

• the function is within the scope of the class,

• the function must be invoked by an object (has a *this* pointer).

A *friend* function of a class is considered as follows:

• the function is a non-member function of the class,

• the function can access all the members (public or private) of the class like the member functions,

• the function declaration can be placed in either the private or the public part of a class declaration.

4.8.1 Friend Functions

The *friend* specifier appears only in the function prototype in the class definition, not in the friend function definition.

The declaration of a friend function in the class is as follows:

Syntax
```
class class_name {
    //...
    friend returnValueType function_name (parameter list);
    //...
};
```

The friend function must be defined outside the class because it is a non-member function of the class.

The syntax of its definition outside the class is as follows:

Syntax
```
returnValueType function_name (parameter list)
{    }
```

Example 4-10: A friend function.

```
//--------------------------------------------------------------------
//File: Date.h
//This program defines a friend function within the Date class.
//--------------------------------------------------------------------
1    class Date{
2    public:
3        Date(int, int, int);
4        //add n to member year
5        friend void add_year(Date&, int);    //declaration of the friend function
6        void print() const;
7    private:
8        int day, month, year;
9    };
//--------------------------------------------------------------------
//File: Date.cpp
//The program presents the implementation detail of the Date class.
//--------------------------------------------------------------------
1    #include "Date.h"
2    #include <iostream>
3    using namespace std;
4
5    Date::Date(int d, int m, int y) : day(d), month(m), year (y) {}
6    void Date::print() const
7    {
8        cout << day << month << year << endl;
9    }
10   //definition of the friend function
11   void add_year(Date& d, int n)
12   {
13       d.year += n;
14   }
//--------------------------------------------------------------------
```

```
//File: example4_10.cpp
//This program demonstrates how to use a friend function.
//----------------------------------------------------------------
1 #include "Date.h"
2 #include <iostream>
3 using namespace std;
4 int main()
5 {
6     Date today(19, 7, 2011);
7     today.print();
7     add_year(today, 5);            //invoke a friend function
8     today.print();
9     return 0;
10 }
```

In Example 4-10, data member *year* is declared in the private part of class *Date*. Friend function *add_year* has direct access to the private member *year* of the class. Thus, a friend function looks like any ordinary functions.

Sometimes a friend function can be a member of another class. For example,

Example 4-11: A friend function as a member of another class.

```
//----------------------------------------------------------------
//File: example4_11.cpp
//The program shows a friend function print in class A, being a member of class B.
//----------------------------------------------------------------
1    #include <iostream>
2    using namespace std;
3
4    class A;
5    class B{
6    public:
7        void print(A& a);
8    };
9    class A{
10   public:
11       A() : x(1), y(2){}
12   private:
13       int x, y;
14       friend void B::print(A& a);      //friend function of class A
15   };
16
17   void B::print(A& a)
18   {
19       cout << "x is " << a.x << endl;
20       cout << "y is " << a.y << endl;
21   }
```

```
22
23    int main()
24    {
25        A aObj;
26        B bObj;
27        bObj.print(aObj);
28        return 0;
29    }
```

Result:

```
1    x is 1
2    y is 2
```

Here the *A* class must be defined before the member function *print* of the *B* class can be a friend of class *A*.

4.8.2 Friend Classes

You can declare an entire class as a friend. Suppose class *B* is a friend of class *A*. This means that every member function and static data member defined in class *B* has access to class *A*.

Example 4-12: A friend class.

```
//---------------------------------------------------------------------------
//File: example4_12.cpp
//The program shows friend class B declared in class A.
//---------------------------------------------------------------------------
1     #include <iostream>
2     using namespace std;
3
4     class A {
5     public:
6         A() : x(1), y(2) {}
7     private:
8         int x, y;
9         friend class B;         //a friend class of class A
10    };
11
12    class B {
13    public:
14        void print(A& a);
15    };
16
17    void B::print(A& a)
18    {
19        cout << "x is " << a.x << endl;
```

```
20        cout << "y is " << a.y << endl;
21    }
22
23    int main()
24    {
25        A aObj;
26        B bObj;
27        bObj.print(aObj);
28        return 0;
29    }
```

Result:

```
1    x is 1
2    y is 2
```

As we can see, the friend class *B* has a member function *print* that accesses the private data members *x* and *y* of class *A* and performs the same task as the friend function *print* in Example 4-11. Any other members declared in class *B* also have access to all members of class *A*.

Think These Over

1. What are the differences among member functions, non-member functions and friend functions?
2. What is the purpose of using friend functions?

4.9 Case Study: Advance of the GradeBook Class

Generally, small, heavily-used abstractions are common in many applications. C++ and other programming languages directly support a few such abstractions. However, most are not, and cannot be, directly supported because there are too many of them. Furthermore, the designer of a general-purpose programming language cannot foresee the detailed needs of every application. Consequently, mechanisms must be provided for the user to define small concrete types. Such types are called concrete types or concrete classes to be distinguished from the abstract classes (see §7.3) and the classes in class hierarchies (see §6.1).

To ensure a user-defined type, to be correct, readable and understandable, a few sets of operations should be considered:

• A constructor specifying how objects of the type are to be initialized and examining the validation of initialized data members.

• A set of functions allowing a user to examine a class object. These functions are marked const to indicate that they do not modify the state of the object for which they are called.

• A set of functions allowing the user to manipulate the object without actually having to know the details of the representation or fiddle with the intricacies of the semantics.

• A set of implicitly defined operations to allow objects to be copied freely.

According to the operations above, a better class can be built.

In Section 3.8, we defined a *GradeBook* class. As we know, a grade book is used to maintain a number of student grades. However, we did not define the attribute of the student grade in this class. Moreover, only a grade attribute is not enough to represent all meanings of a student. Suppose that we consider each student has the attributes of student ID, name and grade. Therefore, a data member we add into the GradeBook class is an aggregation of several built-in types, that is, an object of the *Student* class type is considered as a member of the *GradeBook* class. Thus, "a grade book has a student" is an aggregation relationship between the *GradeBook* class and the *Student* class. The UML diagram of their relationship is shown in Figure 4-5. The empty diamond indicates aggregation relationship in Figure 4-5 (Notice that composition is a special case of the aggregation relationship). The underlined text indicates static class members.

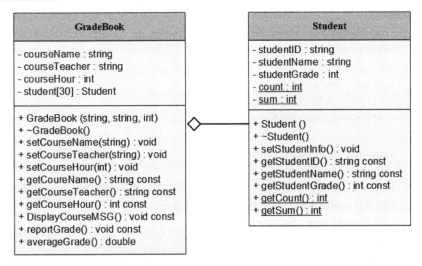

Figure 4-5   The UML diagram of the aggregation relationship

Example 4-13: Demonstrate how to define a *GradeBook* class and use it.

```
//---------------------------------------------------------------------------------
//File: GradeBook.h
//The program presents the declaration of two classes Student and GradeBook.
//---------------------------------------------------------------------------------
1   const int MaxNumberOfStudent = 30;    //the maximum number of students
2   //Student class
3   class Student{
4   public:
5       Student();
```

```cpp
6      ~Student();
7
8      void setStudentInfo();
9      string getStudentID() const;
10     string getStudentName() const;
11     int getStudentGrade() const;
12
13     static int getSum();
14     static int getCount();
15   private:
16     string studentID;           //student ID
17     string studentName;         //student name
18     int studentGrade;           //student grade
19     static int count;           //student number
20     static int sum;             //a sum of student grades
21   };
22
23   //GradeBook class
24   class GradeBook
25   {
26   public:
27     GradeBook(string coursename, string courseteacher, int coursehour, int studentNum);
28     ~GradeBook();
29
30     void setCourseName( string name );
31     void setCourseTeacher( string teacher );
32     void setCourseHour( int hour);
33
34     string getCourseName() const;
35     string getCourseTeacher() const;
36     int getCourseHour() const;
37     //display course messages to the GradeBook user
38     void displayCourseMSG() const;
39     //report all student grades
40     void reportGrade() const;
41     //calculate the average of students
42     double averageGrade();
43   private:
44     string courseName;          //course name
45     string courseTeacher;       //course teacher
46     int courseHour;             //course hour
47     int studentNumber;
48     Student *student;           //a number of students
49   };
//--------------------------------------------------------------------------------
//File: GradeBook.cpp
//The program presents the implementation of classes *Student* and *GradeBook*.
```

```
//-------------------------------------------------------------------------------------------------
1   #include <iostream>
2   #include <string>
3   #include <iomanip>
4   #include <conio.h>
5   using namespace std;
6
7   #include "GradeBook.h"           //include the definition of class GradeBook
8   //initialize the static data members
9   int Student::count = 0;
10  int Student::sum = 0;
11  //implementation detail of the Student class
12  Student::Student()
13  {
14      studentID = " ";   studentName = " ";   studentGrade = 0;
15  }
16  Student::~Student()
17  {
18      if(count > 0)
19          cout << "Destroying student object" << count << endl;
20      --count;
21  }
22  void Student::setStudentInfo()
23  {
24      cout << " Enter student's ID, name and grade\n";
25      cin >> studentID >> studentName >> studentGrade;
26      count ++;   sum += studentGrade;
27  }
28  string Student::getStudentID() const
29  {   return studentID;   }
30  string Student::getStudentName() const
31  {   return studentName;   }
32  int Student::getStudentGrade() const
33  {   return studentGrade;   }
34  int Student::getSum()
35  {   return sum;   }
36  int Student::getCount()
37  {   return count;   }
38  //implementation details of the GradeBook class
39  GradeBook::GradeBook(string coursename, string courseteacher,
40                       int coursehour, int studentNum)
41  {
42      setCourseName( coursename );          //call set function to initialize courseName
43      setCourseTeacher(courseteacher);
44      setCourseHour( coursehour);
45
46      studentNumber = studentNum;
```

```cpp
47        student = new Student[studentNumber];    //declare the student objects;
48
49        int i = 0;
46        while(i < studentNumber)
47        {
48            student[i].setStudentInfo();
49            ++i;
50        }
51   }
52   GradeBook::~GradeBook()
53   {   cout << "Destroying GradeBook object\n";
54       delete []student;    }                    //destroy all student objects
55
56   void GradeBook::setCourseName( string name )
57   {   courseName = name;   }                    //store the course name in the object
58   void GradeBook::setCourseTeacher( string teacher )
59   {   courseTeacher = teacher;   }              //store the course name in the object
60   void GradeBook::setCourseHour( int hour)
61   {   courseHour = hour;   }
62
63   string GradeBook::getCourseName() const
64   {   return courseName;   }                    //return object's courseName
65   string GradeBook::getCourseTeacher() const
66   {   return courseTeacher;   }                 //return object's courseTeacher
67   int GradeBook::getCourseHour() const
68   {   return courseHour;   }                    //return object's courseHour
69
70   void GradeBook::displayCourseMSG() const
71   {
72       //call getCourseName to get the courseName
73       cout << "Welcome to the grade book for " << getCourseName() << "!" << endl;
74       cout << "Teacher: " << getCourseTeacher() << "; "
75            << "Hour: " << getCourseHour() <<endl;
76   }
77   void GradeBook::reportGrade() const
78   {
79       cout << "Student Grade Report\n";
80       cout << setw(10) << "ID" << setw(10) << "Name" << setw(4) << " Grade\n";
81       for(int i = 0; i < Student::getCount(); i++)
82           cout << setw(10) << student[i].getStudentID()
83                << setw(10) << student[i].getStudentName()
84                << setw(4) << student[i].getStudentGrade() <<endl;
85   }
86   double GradeBook::averageGrade()
87   {
88       return (double)Student::getSum() / Student::getCount();
89   }
```

//--
//File: **example4_13.cpp**
//The program tests the implementation of classes *Student* and *GradeBook*.
//--

```
1   #include <iostream>
2   using namespace std;
3   #include "GradeBook.h"
4   int main()
5   {
6      int number;
7      do
8      {
9         cout << "Enter the number of student:";
10        cin >> number;
11     } while (number > MaxNumberOfStudent);
12     //create a GradeBook object
13     GradeBook gradeBook("Object-Oriented Programming in C++", "Christina", 48, number);
14     cout << "--------------------------------------------------------------\n";
15     gradeBook.displayCourseMSG();
16     cout << "--------------------------------------------------------------\n";
17     gradeBook.reportGrade();
18     cout << "--------------------------------------------------------------\n";
19     cout << "The average grade of students: " << gradeBook.averageGrade() << endl;
20     cout << "--------------------------------------------------------------\n";
21     return 0;
22  }
```

Input:

```
Enter the number of student:3
Enter student's ID, name and grade
1002301 wang 75
Enter student's ID, name and grade
1002302 liu 90
Enter student's ID, name and grade
1002301 zhao 65
```

Result:

```
1   --------------------------------------------------------------
2   Welcome to the grade book for Object-Oriented Programming in C++!
3   Teacher: Christina; Hour: 48
4   --------------------------------------------------------------
5   Student Grade Report
6       ID         Name       Grade
7       1002301    wang       75
8       1002302    liu        90
9       1002301    zhao       65
10  --------------------------------------------------------------
```

```
11    The average grade of students: 76.6667
12    --------------------------------------------------------------------------------
13    Destroying GradeBook object
14    Destroying student object3
15    Destroying student object2
16    Destroying student object1
```

Word Tips

alternatively	*adv.*	或者
application	*n.*	应用程序
assignment	*n.*	赋值
composition	*n.*	组合
duplicate	*vt./vi.*	副本，复制
dynamically	*adv.*	动态地
foresee	*vt.*	预知
impractical	*adj.*	不现实的
increment	*n.*	增量
instantiate	*vt.*	举例
intricacy	*n.*	错综复杂
literal	*adj.*	符号的
obscure	*adj.*	费解的
potential	*adj.*	有可能的
recommend	*vt./vi.*	建议，推荐
shallow	*adj.*	浅的
stack	*n.*	栈
storage	*n.*	存储
string	*n.*	字符串
suffix	*n.*	后缀
violate	*vt.*	违反，违背

Exercises

1. Mark the following statements as true(T) or false(F) and give reasons.

(1) Assigning an object to an object of the same type results in default member-wise copy.

(2) Member functions declared const cannot modify the object.

(3) An object cannot be declared as const.

(4) A class cannot have objects of other classes as members.

(5) If a member initializer is not provided for a member object, the member object's default constructor is called implicitly.

(6) Static members are shared by all instances of a class.

(7) A class's static member exists only when an object of the class exists.

(8) The primary activity in C++ is creating objects from the abstract data types and expressing the interactions between those objects.

(9) A class's friend function can access all private data of the class.

(10) A class's friend function is a member of the class.

2. Find the error(s) in each of the following codes and explain possible corrections.

(1)
```
class AA{
private:
    int a;
    int b;
public:
    AA( int x)
    {
        a = 0;
        b = x;
    }
    void print() const
    {   b++;
        cout<<b; }
};
void main()
{   AA   aa;
    aa.print();
    cout<<AA::a;
}
```

(2)
```
class A{
    int x;
    Date d;
    static int y;
public:
    void A(int a)
    {   x = a;
        d(a, a, a)
        y = 0;
    }
};
void main()
{   A aa;
    cout<<aa.y<<endl;
    A bb(10);
    cout<<bb.y<<endl;
}
```

3. Write out the output of the following codes:

(1)
```
class A{
    int a;
public:
    A()
```

```
        {
            a=0;
            cout << "constructing default A" << endl;
        }
        A(int x)
        {
            a = x;
            cout << "constructing A" << endl;
        }
        ~A(){ cout << "destructing A" << endl;}
};
class B{
        A a;
public:
        B()
        {   cout << "constructing default B" << endl;   }
        B(int x):a(x)
        {   cout << "constructing B" << endl;   }
        ~B(){   cout << "destructing B" << endl;   }
};
int main()
{
        B b1;
        B b2(10);
        return 0;
}
```

(2)
```
class Demo{
        int n;
public:
        Demo(){}
        Demo(int m){ n=m; }
        friend Demo square(Demo s){
            Demo s1;
            s1.n = s.n * s.n;
            return s1;
        }
        void disp(){   cout << n << endl;   }
};
void main(void){
        Demo a(10);
        a = square(a);
        a.disp();
}
```

(3)
```
class Demo{
    int n;
    static int sum;
public:
    Demo(int x){   n=x;   }
    void add(){   sum+=n;   }
    void disp(){   cout << "n=" << n << ", sum=" << sum << endl;   }
};
int Demo::sum = 0;
void main(){
    Demo a(2), b(3), c(5);
    a.add();      a.disp();
    b.add();      b.disp();
    c.add();      c.disp();
}
```

(4)
```
class Count{
  public:
      int value;
      Count(const Count& c)
      {  value = c.value;   cout << "Copy an object\n";   }
      Count()
      {  value = 0;   cout << "Creating an object \n";   }
};
void increment (Count c, int time)
{    cout << "Calling the increment function\n";
     c.value++;
     time++;   }
int main()
{
    Count myCount;
    int times = 0;
    for (int i = 0; i < 3; i++)
        increment(myCount, times);
    cout << " myCount.count is " << myCount.value << endl;
    cout << " times is " << times << endl;
    return 0;
}
```

If the statement *void increment(Count c, int times)* is changed to

   void increment(Count& c, int& times),

what will be the output?

4. Write a program that the *Cube* class is defined with the properties *x, y, z, length, width,*

· 135 ·

and *height*. The class includes the following definition:

(1) two constructors, a copy constructor and a destructor;

(2) a member function to calculate the volume of *Cube*;

(3) a member function to display the properties;

(4) a member function to move a *Cube*;

(5) a member function to modify the properties of *Cube*.

Test the *Cube* class by using the main function.

5. Define a class called *Employee* that contains a *name* (an object of the *string* type) and an employee *ID* (*long* type). Include a member function called *getdata* to get data from the user for insertion into the object, and another function called *putdata* to display the data. Assume the name has no embedded blanks. Write a *main* function to execute this class. It should create an array of type *Employee*, and then invite the user to input data for up to 100 employees. Finally, it should print out the data for all the employees.

6. Define a *Dictionary* class that has the *Word* objects defined in the Exercises section of Chapter 3 and the number of words. Include member function *FindWord* to find a word, function *AddWord* to add a word, function *GetWord* to obtain a word and function *Print* to output all words.

7. Write the definition of three objects *Bank*1, *Bank*2 and *Bank*3. These three objects all have private data member *balance* and a member function *display* to output the balance. Design a friend function *total* in the three objects to calculate the total balance from these banks. Test the program in the *main* function.

8. Design a class named *MyInteger*. Draw the UML diagram for the class. Implement the class. Write a test program that tests all functions in the class.

The class contains the following:

(1) an *int* data member named *value* that stores the *int* value.

(2) a constructor that creates a *MyInteger* object for the specified *int* value.

(3) a constant *get* function that returns the *int* value.

(4) a constant function *isEven()* that returns true if the value is even.

(5) a static function *isEven(int)* that returns true if the specified value is even.

(6) a constant function *isEqual(const MyInteger&)* that returns true if the value in the object is equal to the specified object's value.

(7) a function *parseInt(const string&)* that converts a *string* to an *int* value.

Chapter 5 Operator Overloading

When I use a word it means just what I choose it to mean—

neither more nor less

—Humpty Dumpty

Objectives
- To learn about operator overloading
- To explore the member and non-member operator functions of a class
- To understand how to overload operators to work with user-defined types
- To understand how to convert objects from one class to another class

5.1 Introduction to Operator Overloading

The previous chapter defined and implemented the *Date* class. It also showed how the *Date* class can be used to represent a date in the user's program. Let's review the definition of the *Date* class.

Two objects are declared as follows:

```
Date myDay(12, 6, 2000);
Date yourDay(23, 9, 2005);
```

Now we do the following operations:

```
myDay.print();
myDay.add_year(1);
if(myDay.equalTo(yourDay)
...
```

The first statement prints the value of the *myDay* object. Whether can we use the insertion operator << to output the value of *myDay*? The second statement increments the value of *myDay* by one year, whether can we use the increment operator ++ to increment the value of *myDay*? Likewise, whether can we use a relational operator for comparison? If yes, we can enhance the flexibility for manipulating object *myDay*. Thus, we can use the following statements instead of the previous statements:

```
cout<<myDay;
myDay++;
if (myDay == yourDay)
```

As we know, C++ supports a set of operators for built-in types. However, these operators cannot be applied directly to user-defined types. Although we want to use the operators <<, ++ and == to represent the notions of the *myDay* object, this will cause compile errors since we do not define the behavior of the *Date* class with <<, ++ and ==. Therefore, we must extend the definition of these operators in our class *Date*. This is called ***operator overloading***, in C++ terminology. Operator overloading is one of the most exciting features of object-oriented programming.

5.2 Operator Functions

5.2.1 Overloaded Operators

C++ allows the user to overload most of the operators so that the operations can work effectively in a specific application. Here is a list of all the operators that can be overloaded:

+	-	*	/	=	<	>	+=	-=	*=	/=
<<	>>	<<=	>>=	==	!=	<=	>=	++	--	
%	&	^	!	|	~	&=	^=	|=	&&	||
%=	[]	()	->*	->	new	delete	new[]	delete[]		

Actually, C++ does not allow the users to overload all operators. The following operators cannot be defined by a user:

::   (scope resolution),
.   (member selection),
.*   (member selection through pointer to member),
?:   (ternary operator)

5.2.2 Operator Functions

In order to overload operators, you must write functions. The function that overloads an operator is named by the reserved word ***operator*** followed by the operator to be overloaded.

> The **operator function** is a function that overloads an operator.
> 运算符函数是一个重载运算符的函数。

The format of the operator function is

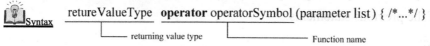

    retureValueType **operator** operatorSymbol (parameter list) { /*...*/ }
                   └── returning value type   └── Function name

First, we give an example of the class complex number. A number of the form *a + ib*, where $i^2 = -1$, and *a* and *b* are real numbers, is called a ***complex number***.

Example 5-1: A member operator function and a non-member operator function.

```
//-----------------------------------------------------------------
//File: complex.h
//This program describes the definition of a complex class in which there is a member
//operator (+) function.
//-----------------------------------------------------------------
1    #include <iostream>
2    using namespace std;
3
4    class complex{
5    public:
6        complex(double, double);
7        //operator function as a member
8        complex operator + (complex&);           //overloading operator +
9        void print() const;
10
11       double real, imag;
12   };
//-----------------------------------------------------------------
//File: complex.cpp
//The program presents the implementation detail of the complex class and the definition
//of a non-member operator (-) function.
//-----------------------------------------------------------------
1    #include "complex.h"
2
3    complex::complex(double re, double im) : real( re ), imag( im ){}
4
5    complex complex::operator + (complex& c)     //overloading operator+
6    {
7        complex temp(0, 0);
8        temp.real = real + c.real;
9        temp.imag = imag + c.imag;
10       return temp;
11   }
12   void complex::print() const
13   {
14       cout << real<< " + " << imag << "i\n";
15   }
16   //operator function as a non-member
17   complex operator - (complex& c1, complex& c2)   //overloading operator-
18   {
19       complex temp(0, 0);
20       temp.real = c1.real - c2.real;
21       temp.imag = c1.imag - c2.imag;
22       return temp;
```

```
        23    }
//------------------------------------------------------------------------------
//File: example5_1.cpp
//This program shows how to overload operators + and -.
//------------------------------------------------------------------------------
        1     #include "complex.h"
        2
        3     int main()
        4     {
        5         complex x(2.3, 4.5), y(3.0, 1.2);
        6         complex z = x + y;        //shorthand: equivalent to z=x.operator + (y)
        7         z.print();
        8         z = x – y;                //z = operator - (x, y)
        9         z.print();
        10        return 0;
        11    }
```

Result:

```
        1     5.3 + 5.7i
        2     -0.7 + 3.3i
```

Example 5-1 presents two operator functions of addition operator (+) and minus operator (-). They are defined as follows:

```
complex complex::operator + (complex& c)
complex operator - (complex& c1, complex& c2)
```

The first function overloads an addition operator (+) as a member function. The second function overloads a minus operator (-) as a non-member function. Notice that we have to access data members *real* and *imag* in the non-member operator function so we put these data members into the *public* part of the class declaration in **complex.h**.

However, putting data members *real* and *imag* into the public part of class *complex* violates the information hiding of the class. We can define a minus operator (-) as a friend function.

Example 5-2: A friend operator function.

```
//------------------------------------------------------------------------------
//File: complex.h
//This program describes the definition of a complex class in which there is a friend
//operator (-) function.
//------------------------------------------------------------------------------
        1     #include <iostream>
        2     using namespace std;
        3
        4     class complex{
        5     public:
```

```
6      complex(double, double);
7      //operator function
8      friend complex operator - (complex&, complex&);
9      void print() const;
10  private:
11     double real,  imag;
12  };
```
//--
//File: **complex.cpp**
//The program presents the implementation detail of the *complex* class and the definition
//of a friend operator (-) function.
//--
```
1   #include "complex.h"
2
3   complex::complex(double re, double im) : real( re ), imag( im ){}
4
5   void complex::print() const
6   {
7       cout << real << " + " << imag << "i\n";
8   }
9   //definition of the friend function
10  complex operator - (complex& c1, complex& c2)
11  {
12      complex temp(0, 0);
13      temp.real = c1.real - c2.real;
14      temp.imag = c1.imag - c2.imag;
15      return temp;
16  }
```

The user program is the same as that of Example 5-1. In Example 5-2, data members *real* and *imag* are declared in the private part of class *complex*. Friend function *operator* - has direct access to these private members of the class. Thus, a friend function is an ordinary function.

The above gives an example of that the + and - operators are overloaded as a member function and non-member function. So, when can the operators be overloaded as either member functions or non-member functions?

As we know, the + operator is overloaded as a member function. The + operator has direct access to the data members of the objects, and you need to pass only one object as a parameter. On the other hand, when the - operator is overloaded as a non-member function, then you must pass both objects as parameters. Therefore, overloading operators as a non-member could require additional memory and computational time to make a local copy of the data. Thus, for efficiency purpose, wherever possible, you should overload operators as member functions.

When overloading an operator, keep the following in mind:
- You cannot change the precedence of an operator.
- The associativity cannot be changed. For example, the associativity of arithmetic operator addition is from left to right, and it cannot be changed.

- Default parameters cannot be used with an overloaded operator.
- You cannot change the number of parameters an operator takes.
- You cannot create new operators. Only existing operators can be overloaded.
- The following operators cannot be overloaded:

    ::   .   .*   ?:   sizeof(), typeid()

- The meaning of how an operator works with built-in types, such as int, remains the same.
- Operators can be overloaded either for objects of the user-defined types, or for a combination of objects of the user-defined types and objects of the built-in types.

当重载运算符时，请记住以下几点：
- 不能改变运算符的优先级；
- 不能改变运算符的结合性。例如，算术运算符加法是从左到右结合，并且不能改变；
- 默认参数不能与重载操作符一起使用；
- 不能改变运算符所需的参数数目；
- 不能创建新的运算符。只有现有运算符可以重载；
- 下列运算符不能重载：

    ::   .   .*   ?:   sizeof(),   typeid()

- 运算符如何与内置类型（如 int）一起运算的含义保持不变；
- 运算符可以重载或为用户定义的类型的对象，或一个用户定义的类型和内置类型组合的对象。

 Think These Over
1. When do we need to overload the operators?
2. When can the operators be overloaded as either member functions or non-member functions?

5.3 Binary and Unary Operators

5.3.1 Overloading Binary Operators

In general, arithmetic, such as +, -, *, or relation, such as ==, <=, represent a binary operator. These ***binary operators*** can be defined by either a non-static *member* function taking one parameter or a *non-member* (or *friend*) function taking two parameters.

Overloading Binary Operators as Member Functions

If the operator function is a member of the class, it has one parameter.

The declaration of its prototype is

Syntax classname operator@(classname&);

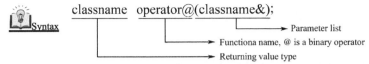

The syntax of its definition outside the class is

Syntax classname classname::operator@ (classname&);

For example,

```
1    class complex{
2    public:
3        complex(double, double);
4
5        //overloading a binary operator +
6        complex operator + (complex&);     //member function taking one parameter
7    private:
8        double real, imag;
9    };
```

Overloading Binary Operators as Non-Member Functions

If the operator function is a non-member (or friend) of the class, it has two parameters.

The syntax of its definition is

Syntax classname operator@ (classname&, classname&);

For example,

```
1    //non-member function
2    complex operator - (complex& c1, complex& c2)   //taking two parameters
3    {
4        complex temp(0, 0);
5        temp.real = c1.real - c2.real;
6        temp.imag = c1.imag - c2.imag;
7        return temp;
8    }
```

5.3.2 Overloading Unary Operators

The process of overloading unary operators is similar to the process of overloading binary operators. The only difference is that in the case of unary operators, the operator has only one parameter.

Overloading Unary Operators as Member Functions

If the operator function is a member of the class, it has no parameter.

The declaration of its prototype is

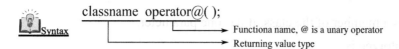

Syntax: classname operator@( );
— Functiona name, @ is a unary operator
— Returning value type

The syntax of its definition outside the class is

Syntax   classname classname::operator@ ();

Overloading Unary Operators as Non-Member Functions

If the operator function is a non-member (or friend), it has one parameter.

The syntax of its definition is

Syntax   classname operator@ (classname&);

Now we describe how to overload two unary operators, that is, increment (++) and decrement (--) operators.

The increment operator has two forms: ***prefix*** increment (++a) and ***postfix*** increment (a++), where *a* is a type variable, such as an *int* type, or a user-defined type. In the case of prefix increment, the value of *a* is incremented by 1 before the value is used in the expression. In the case of postfix increment, the value of *a* is used in the expression before it is incremented by 1.

Overloading Prefix Increment Operator (++)

Suppose that *c* is an object of class *complex*, the statement

    ++c;

increments the values of *real* and *imag* of *complex* by 1.

Thus, overloading an increment operator (++) as a member of the *complex* class is defined as follows:

```
1   class complex{
2   public:
3       complex(double, double);
4
5       //overloading a unary ++ operator as member function
6       complex& operator ++ ();          //member function taking no parameter
7   private:
8       double real, imag;
9   };
10  complex& complex::operator++()        //definition of member function
11  {
12      real++;
13      imag++;
14      return *this;
15  }
```

The following example is that overloading the ++ operator as a friend of the *complex* class.

```
1    class complex{
2    public:
3        complex(double, double);
4        //overloading a unary ++ operator as non-member function
5        friend complex operator ++ (complex&);      //taking one parameter
6    private:
7        double real, imag;
8    };
9
10   complex operator++(complex& c)            //definition of friend function
11   {
12       (c.real)++;
13       (c.imag)++;
14       return c;
15   }
```

Overloading Postfix Increment Operator

The examples above present the prefix increment operator (++) to be overloaded. To distinguish between pre- and post-fix increment operator overloading, we use a dummy parameter (of type *int*) in the function declaration of the operator function. Thus, the function prototype for the postfix increment operator of the *complex* class is:

```
complex& operator++(int)              //for the member function, or
complex operator++(complex&, int)     //for the non-member function (or friend)
```

Suppose that *c* is an object of class *complex*, the statement:

```
c++;
```

is compiled by compiler in the statement: *c.operator++(0)*.

Example 5-3: The *complex* class with overloading prefix and postfix operator (++).

```
//----------------------------------------------------------------------
//File: complex.h
//This program describes the definition of a complex class with the friend function of prefix
//increment operator and a member function of postfix increment operator.
//----------------------------------------------------------------------
1    #include<iostream>
2    using namespace std;
3    class complex{
4    public:
5        complex(double = 0, double = 0);
6
7        //non-member function
8        friend complex operator++ (complex&);     //prefix increment operator
```

· 145 ·

```
9
10      //member function
11      complex operator++ (int);                    //postfix increment operator
12      void print() const;
13  private:
14      double real, imag;
15  };
```
//--
//File: **complex.cpp**
//The program presents the implementation detail of the *complex* class.
//--
```
1   #include "complex.h"
2
3   complex::complex(double re, double im) : real(re), imag(im) {}
4   //member function
5   complex complex::operator++ (int)
6   {
7       complex temp = *this;
8       real++;
9       imag++;
10      return temp;
11  }
12  //non-member function
13  complex operator++ (complex& c)
14  {
15      c.real++;
16      c.imag++;
17      return c;
18  }
19  void complex::print() const
20  {
21      cout << real << " + " << imag << "i\n";
22  }
```
//--
//File: **example5_3.cpp**
//This program shows how to overload prefix and postfix operators ++.
//--
```
1   #include "complex.h"
2
3   int main()
4   {
5       complex c1(20.3,11.3);
6
7       cout << "Original value of c1: ";
8       c1.print();
9
10      ++c1;                        //prefix increment
```

· 146 ·

```
11      cout << "Value after ++c1: ";
12      c1.print();
13
14      c1++;                          //postfix increment
15      cout << "Value after c1++: ";
16      c1.print();
17
18      complex c2 = ++c1;
19      cout << "\nValue of c1 after c2 = ++c1:";
20      c1.print();
21      cout << "Value of c2 after c2 = ++c1:";
22      c2.print();
23
24      c2 = c1++;
25      cout << "\nValue of c1 after c2 = c1++:";
26      c1.print();
27      cout << "Value of c2 after c2 = c1++:";
28      c2.print();
29      return 0;
30    }
```

Result:

```
1   Original value of c1: 20.3 + 11.3i
2   Value after ++c1: 21.3 + 12.3i
3   Value after c1++: 22.3 + 13.3i
4
5   Value of c1 after c2 = ++c1: 23.3 + 14.3i
6   Value of c2 after c2 = ++c1: 23.3 + 14.3i
7
8   Value of c1 after c2 = c1++: 24.3 + 15.3i
9   Value of c2 after c2 = c1++: 23.3 + 14.3i
```

Think It Over

Why do the operator functions take different parameters between member and non-member functions?

5.4 Overloading Combinatorial Operators

The meanings of some built-in operators are defined to be equivalent to some combination of other operators on the same arguments. For example, if *a* and *n* are integers, *a* += *n* means *a* = *a* + *n*. Such relations do not hold for user-defined operators unless the user happens to define them that way. For example, a compiler would not generate a definition of *complex::operator+=()* from the definitions of *complex::operator+()* and *complex::operator = ()*.

· 147 ·

Therefore, you must predefine an addition assignment operator (+=).

Example 5-4: Overloading a combinatorial operator (+=).

```
//----------------------------------------------------------------
//File: complex.h
//This program describes the definition of the complex class with overloading operator +=.
//----------------------------------------------------------------
1    #include<iostream>
2    using namespace std;
3
4    class complex{
5    public:
6        complex(double = 0, double = 0);      //constructor
7        complex(const complex&);              //copy constructor
8
9        //member function
10       complex operator+ (complex&);         //overloading addition +
11       complex& operator= (complex&);        //overloading assignment =
12
13       complex& operator+= (complex&);       //predefining operator +=
14       void print() const;
15   private:
16       double real, imag;
17   };
//----------------------------------------------------------------
//File: complex.cpp
//The program presents the implementation detail of the complex class.
//----------------------------------------------------------------
1    #include "complex.h"
2
3    complex::complex(double re, double im) : real(re), imag(im)
4    {
5        cout << "Constructing an object\n";
6    }
7    complex::complex(const complex& c)
8    {
9        real = c.real;
10       imag = c.imag;
11       cout << "Copying an object\n";
12   }
13
14   complex complex::operator+ (complex& c)
15   {
16       complex temp;
17       temp.real = real + c.real;
18       temp.imag = imag + c.imag;
19       return temp;
```

· 148 ·

```
20      }
21      complex& complex::operator= (complex& c)
22      {
23              real = c.real;
24              imag = c.imag;
25              cout << "Overloading assignment operator =\n";
26              return *this;
27      }
28      complex& complex::operator+= (complex& c)
29      {
30              *this = *this + c;
31              return *this;
32      }
33      void complex::print() const
34      {
35              cout << real << " + " << imag << "i\n";
36      }
```

//--
//File: **example5_4.cpp**
//This program shows how to overload operator +=.
//--

```
1       #include "complex.h"
2
3       int main()
4       {
5               complex c1(20.3,11.3);
6               cout << "Original value of c1: ";
7               c1.print();
8
9               complex c2 = c1;
10              cout << "Original value of c2: ";
11              c2.print();
12
13              complex c3;
14              cout << "value of c3 after c3 = c1 is \n";
15              c3 = c1;
16              c3.print();
17              cout << "value of c3 after c3 += c2 is ";
18              c3 += c2;
19              c3.print();
20              return 0;
21      }
```

Result:

1	Constructing an object
2	Original value of c1: 20.3 + 11.3i
3	Copying an object

```
4  Original value of c2: 20.3 + 11.3i
5  Constructing an object
6  value of c3 after c3 = c1 is
7  Overloading assignment operator =
8  20.3 + 11.3i
9  value of c3 after c3 += c2 is 40.6 + 22.6i
```

You may notice that the two statements in Lines 9 and 15 (in example5_4.cpp) output the different results. These two statements look very similar. However, they have different meanings. The statement in Line 9 initializes an object $c2$ by the constructed object $c1$ and invokes the copy constructor of class *complex*, doing a member-wise copy of object $c1$. The output is shown in Line 3 of the Result. The statement in Line 15 invokes the assignment operator (=) function of class *complex*, assigning object $c1$ to object $c2$. The output is shown in Line 7 of the Result.

Copy Constructor and Assignment Operator

Sometimes, the default copy constructor and default assignment operator are used in the program. If we do not define a copy constructor and an assignment operator in the file complex.h of Example 5-4, we use the following statements:

```
complex c2 = ++c1;
c2 = c1++;
```

The first statement invokes a default copy constructor to initialize the $c2$ object by $++c1$. The second statement invokes a default assignment operator to assign the new data from $c1++$ to $c2$. The situation with the default implementations is that a simple copy of the members (i.e. **Shallow copy**) may not be appropriate to clone an object.

For instance, what if one of the members was a pointer that is allocated by the class? The shallow copying the pointer is not enough because now you'll have two objects that have the same pointer value, and both objects will try to free the memory allocated to that pointer when they are destroyed. In this case, we must define a copy constructor and an assignment operator for copying objects (i.e. ***Deep Copy***)

The difference between **copy constructor** and **assignment operator** is that the copy constructor of the target is invoked when the source object is passed in at the time the target is constructed whereas the assignment operator is invoked when the target object already exists.

拷贝构造函数和赋值运算符的区别是，在以传递源对象形式构造目标对象时调用拷贝构造函数，而当目标对象已经存在时则调用赋值运算符。

5.5 Mixed Arithmetic of User-Defined Types

An operator function must either be a member or take at least one argument of a user-defined type (this is not needed for the functions redefining the ***new*** and ***delete*** operators). This rule ensures that a user cannot change the meaning of an expression unless the expression contains an object of a user-defined type. In particular, it is not possible to define an operator function that operates exclusively on pointers. This ensures that C++ is extensible but not mutable (with the exception of operators = and &, for class objects).

Sometimes. an operator function intends to accept a built-in type as an operand. For example,

```
Complex c(2.4, 3.5);
Complex x = c + 2;
Complex y = 2 + c;
```

The second line states that a complex object c is added to the integer 2: $c+2$ can, with a suitably declared member function, which can be interpreted as $c.operator + (2)$, that is,

$$c+2 \iff c.operator + (2)$$

In this case, the operator function is defined as a member function as follows:

```
complex& complex::operator+ (int a)
{
    real += a;
    return *this;
}
```

However, the $2+c$ statement in the third line cannot be done because there is no class *int* for which the definition of operator + means $2.operator + (c)$, that is,

$$2+c \not\iff 2.operator + (c)$$

In this case, the operator function needs to be defined as a non-member function as follows:

```
complex operator+ (int a, complex& c)
{
    return complex(a+c.real, c.imag);
}
```

Even if there were, two different member functions would be needed to cope with $2+c$ and $c+2$. Because the compiler does not know the meaning of a user-defined +, it cannot assume that it is commutative and then interpret $2+c$ as $c+2$.

5.6 Type Conversion of User-Defined Types

Most programs process a variety of types. Sometimes all the operations "stay within a type". For instance, adding an integer to an integer produces an integer. But it is often necessary to convert data of one type to data of another type. This can happen in assignments, in calculations, in passing values to functions, and in returning values from functions. The compiler knows how to perform certain conversion among built-in types. Programmers can force conversions among built-in types by casting.

But what about user-defined types? The compiler cannot automatically know how to convert between user-defined types and built-in types.

5.6.1 Converting a Built-In Type to a User-Defined Type

Such conversion can be performed by a constructor, called ***conversion constructors***, namely, single argument constructors that convert objects of built-in types into objects of a specified class.

Example 5-5: The conversion of a *double* type to a *complex* type.

```
//----------------------------------------------------------------
//File: complex.h
//This program describes the definition of a complex class with conversion constructor.
//----------------------------------------------------------------
1    #include <iostream>
2    using namespace std;
3    class complex{
4    public:
5        complex(double r, double i):real(r), imag(i){}
6        complex(double);         //conversion constructor
7        void print() const;
8    private:
9        double real, imag;
10   };
//----------------------------------------------------------------
//File: complex.cpp
//The program presents the implementation detail of the complex class.
//----------------------------------------------------------------
1    #include "complex.h"
2    complex::complex(double r)
3    {
4        real = r;   imag = 0.0;
5    }
6    void complex::print() const
```

```
7     {
8         cout << real << "+" << imag << "i\n";
9     }
```
//--
//File: **example5_5.cpp**
//This program tests how to convert a *double* type to a *complex* type.
//--
```
1   #include "complex.h"
2   int main()
3   {
4       complex c1(20.3, 11.3);
5       complex c2 = 10.5;
6       c1.print();
7       c2.print();
8       return 0;
9   }
```

Result:

```
1   20.3 + 11.3i
2   10.5 + 0.0i
```

In Line 5 of the *main* function, 10.5 is first converted to an unnamed complex object by invoking the conversion constructor, i.e. *complex(10.5)*, and the *c*2 object is then created by the unnamed object.

Using a constructor to specify type conversion is convenient but has implications that can be undesirable. A constructor cannot specify

(1) an implicit conversion from a user-defined type to a built-in type (because the built-in types are not classes), or

(2) a conversion from a new class to a previously defined class (without modifying the declaration for the old class).

5.6.2 Converting a User-Defined Type to a Built-In Type

When you convert a user-defined type to a built-in type, you can use a conversion operator.

A **conversion operator** (also called a **cast operator**) can be used to convert an object of one class into an object of another class or into an object of a built-in type.

转换运算符（也称为 cast 运算符）可用于将一个类的对象转换为另一个类的对象或转换为内置类型。

Such a conversion operator must be a non-static member function, but not a friend function.

The function of the conversion operator is as follows:

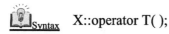 Syntax   X::operator T( );

where *T* is a type name. The statement defines a conversion from a user-defined type *X* to *T*.

Example 5-6: The conversion of a *complex* type to *double* and *int* types.

```
//------------------------------------------------------------------------
//File: complex.h
//This program describes the definition of a complex class with type conversion.
//------------------------------------------------------------------------
1    class complex{
2        double real, imag;
3    public:
4        complex(double =0, double =0);
5        operator double();              //conversion operator
6        operator int();                 //conversion operator
7    };
//------------------------------------------------------------------------
//File: complex.cpp
//The program presents the implementation detail of the complex class.
//------------------------------------------------------------------------
1    #include "complex.h"
2
3    complex::complex(double re, double im) : real(re), imag(im) {}
4
5    complex::operator double()
6    {   return real;    }
7
8    complex::operator int()
9    {   return (int)(real);   }
10
//------------------------------------------------------------------------
//File: example5_6.cpp
//This program illustrates how to convert a complex type to int and double types.
//------------------------------------------------------------------------
1    #include "complex.h"
2    #include <iostream>
3    using namespace std;
4
5    void main()
6    {
7        complex a(2.1,5.6);
8        cout << double(a) << endl;
9        cout << int(a) <<endl;
10   }
```

Result:

```
1    2.1
2    2
```

In Example 5-6, the statements *operator double()* and *operator int()* in Lines 5 and 6 in complex.h mean that the objects of a *complex* type can be converted to the *double* and *int* types.

Conversion operator is different from an ordinary overloaded operator:

(1) It should not return a value (not even void).

(2) It takes no arguments.

转换运算符不同于一般的重载的运算符：

(1) 它不应该返回值（甚至不能写 void 值）。

(2) 它不能有参数。

5.7 Case Study: A MyInteger Class

A *MyInteger* class has the property of an integer value. Define a *MyInteger* class according to the following test program:

```
1   int main()
2   {
3       MyInteger a1;
4       cin >> a1;
5       MyInteger a2 = 5;
6       MyInteger a3 = a1 + a2;
7
8       cout << 2 + a3 << endl;
9       cout << --a3;
10      cout << a2++;
11
12      a1 += a2;
13      if (a1 == a3)
14          cout << "These two objects are same\n";
15      else
16          cout << "These two objects are different\n";
17
18      int sum = 0;
19      for (int i = 0; i < 10; i++)
20      {
21          MyInteger m(i);
22          sum += int(m);
23      }
24      cout << sum << endl;
25      return 0;
26  }
```

Here, the *MyInteger* class is a concrete type. In the previous sections, we present the definitions of operator functions. A complete class design follows the guidelines from Chapter 4. In addition, users of complex arithmetic rely so heavily on those operators that the definition of *MyInteger* brings into play most of the basic rules for operator overloading. Therefore, we will consider the following things in the *MyInteger* class:

(1) Constructors and Copy Constructors

To cope with assignments and initialization of the *MyInteger* object with scalars, we need to convert a scalar (an integer) to a *MyInteger*. For example:

```
MyInteger a1;
MyInteger a2 = 5;   //creating a MyInteger(5) and then a2 is initialized by MyInteger(5)
```

(2) Member and Non-Member Functions

In general, minimizing the number of functions can directly manipulate the representation of an object. This can be achieved by defining only operators that inherently modify the value of their first argument, such as +=, in the class itself. Operators that simply produce a new value based on the values of its arguments, such as +, are then defined outside the class and use the essential operators in their implementation. For example,

```
MyInteger a3 = a1 + a2      //a3= a1.operator(a2)
a1 += a2;                   //a1.operator+=(a2)
if (a1 == a3)               //a1.operator==(a3)
```

(3) Conversion for Mixed Arithmetic

To copy with mixed arithmetic, for example,

```
cout << 2 + a3 << endl;
```

(4) Type Conversion

To copy with type conversion, for example,

```
MyInteger a2 = 5;           //int type converts to MyInteger type
MyInteger m(i);
sum += int(m);              //MyInteger converts to int type
```

(5) Overloading Operators >> and <<

You can use operator >> and << to input/output an object. For example,

```
cin >> a1;
cout << a2++;
```

Now we present the definition of a *MyInteger* class in Example 5-7.

Example 5-7: A *MyInteger* class.

```
//---------------------------------------------------------------
//File: MyInteger.h
```

//This program describes the definition of a *MyInteger* class.
//--
```
1   #include <iostream>
2   using namespace std;
3
4   class MyInteger{
5   public:
6       MyInteger(int = 0);                             //constructor
7       MyInteger(const MyInteger&);                    //copy constructor
8
9       MyInteger operator+(MyInteger&);                //overloading operator +
10      MyInteger& operator+=(MyInteger&);              //overloading operator +=
11      MyInteger& operator--();                        //overloading prefix operator --
12      bool operator==(MyInteger&);                    //overloading operator ==
13      MyInteger operator++(int);                      //overloading postfix operator ++
14      friend MyInteger operator+(int, MyInteger&);    //mixed arithmetic
15
16      operator int();                                 //conversion operator
17
18      friend ostream& operator<<(ostream&, MyInteger&);   //overloading operator <<
19      friend istream& operator>>(istream&, MyInteger&);   //overloading operator >>
20  private:
21      int value;
22  };
```
//--
//File: **MyInteger.cpp**
//The program gives the definitions of the functions to implement various operations of the
//*MyInteger* class.
//--
```
1   #include "MyInteger.h"
2   MyInteger::MyInteger(int m)
3   {
4       value = m;
5       cout << "Constructing an object " << value << endl;
6   }
7   MyInteger::MyInteger(const MyInteger& m)
8   {
9       value = m.value;
10      cout << "Copying an object " << value << endl;
11  }
12  MyInteger MyInteger::operator+(MyInteger& m)
13  {
14      cout << "operator+\n";
15      MyInteger temp;
16      temp.value = value + m.value;
17      return temp;
18  }
```

```cpp
19   MyInteger& MyInteger::operator+=(MyInteger& m)
20   {
21      cout << "operator+=\n";
22      value += m.value;
23      return *this;
24   }
25   MyInteger& MyInteger::operator--()
26   {
27      --value;
28      return *this;
29   }
30   MyInteger MyInteger::operator++(int)
31   {
32      MyInteger m = *this;
33      value++;
34      return m;
35   }
36   bool MyInteger::operator==(MyInteger& m)
37   {
38      if (value == m.value)
39         return true;
40      else
41         return false;
42   }
43   MyInteger operator+(int i, MyInteger& m)
44   {
45      return MyInteger(i + m.value);
46   }
47   MyInteger::operator int()
48   {
49      return value;
50   }
51
52   ostream& operator<<(ostream& output, MyInteger& m)
53   {
54      output << m.value << endl;
55      return output;
56   }
57   istream& operator>>(istream& input, MyInteger& m)
58   {
59      input >> m.value;
60      return input;
61   }
```

According to the given test program, the program has the following output:

Result:

```
1   Constructing an object 0
2   3
3   Constructing an object 5
4   operator+
5   Constructing an object 0
6   Copying an object 8
7   Constructing an object 10
8   10
9   7
10  Copying an object 5
11  Copying an object 5
12  5operator+=
13  These two objects are different
14  Constructing an object 0
15  Constructing an object 1
16  Constructing an object 2
17  Constructing an object 3
18  Constructing an object 4
19  Constructing an object 5
20  Constructing an object 6
21  Constructing an object 7
22  Constructing an object 8
23  Constructing an object 9
24  45
```

Most operator functions can be either member functions or non-member functions (or friend). To make an operator function be a member or non-member function of a class, please keep the following in mind:

(1) The function that overloads any of the operators (), [], ->, or = for a class must be declared as a **member function**.

(2) Suppose that an operator *op* is overloaded for a class called *myClass*.

• If the leftmost operand of *op* is an object of a different class (that is, not of type *myClass*), the function to overload the operator *op* must be a ***non-member*** (or friend) of *myClass*.

• If an operator function to overload the operator *op* for *myClass* is a member of *myClass*, then when applying *op* on objects of *myClass*, the leftmost operand of *op* must be of type *myClass*.

大多数运算符函数既可以是成员函数也可以是非成员函数（或友元函数）。要使运算符函数成为类的成员或非成员函数，请记住以下几点：

（1）一个类重载任何一个运算符（ ）、[]、-> 或 =的运算符函数必须声明为**成员函数**。

（2）假如在一个被称为 myClass 的类中运算符 op 被重载。

• 如果 op 最左侧的操作数不是该类的对象（不是 myClass 类型），则重载运算符 op 的函数必须是 myClass 类的**非成员**（或友元）函数。

- 对于类 myClass，如果重载运算符 op 的运算符函数是 myClass 类的一个成员，那么当 myClass 对象使用运算符 op 时，op 最左侧的操作数必须是 myClass 类型的对象。

Word Tips

allocated *adj.* 分配的		logically *adv.* 理论上地
application *n.* 应用程序		manipulate *vt.* 操作
associativity *n.* 结合性		minimized *adj.* 最少的
binary *adj.* 二元的		mutable *adj.* 多变的
certain *adj.* 确定的		parameter *n.* 参数
combination *n.* 组合		postfix *n.* 后缀
compiling *n.* 编译		precedence *n.* 优先权
complex number *n.* 复数		prefix *n.* 前缀
components *n.* 成分		previous *adj.* 以前的
concept *n.* 概念		prototype *n.* 原型
concrete *adj.* 具体的，实际的		representation *n.* 表示
convert *vt./vi* 转换		respectively *adv.* 分别地，各自地
dynamic *adj.* 动态的		scalar *n.* 标量
effectively *adv.* 有效地		section *n.* 章节
exclusively *adv.* 专门地		store *v.* 储存
explicitly *adv.* 明确地		terminate *v.* 终止
extensible *adj.* 可展开的		terminology *n.* 术语
flexibility *n.* 灵活性		trivially *adv.* 平凡地
guidelines *n.* 指南		unary *adj.* 一元的
increment operator *n.* 增量运算符		undesirable *adj.* 不合意的
library *n.* 库		

Exercises

1. Mark the following statements as true (T) or false (F) and give reasons.

(1) C++ allows operators to be overloaded.

(2) C++ allows new operators to be created.

(3) Any operator can be overloaded.

(4) How an operator works on basic types cannot be changed by operator overloading.

(5) The associativity of an operator can be changed by operator overloading.

(6) The precedence of an operator cannot be changed by operator overloading.

(7) Overloaded assignment operators must be declared as non-member functions.

(8) Overloaded operator member functions for binary operators implicitly use this to obtain access to the left operand of the operator.

(9) A unary operator can be overloaded as a non-member function.

(10) A binary operator can be overloaded as a non-member function taking one argument.

(11) A conversion operator converts an object of one class into an object of another class.

(12) The ++ (or --) operators can be overloaded for both prefix and postfix increment (or decrement).

2. Given the following program.

```cpp
class Rectangle{
private:
    int length;
    int width;
public:
    Rectangle(int l = 0, int w = 0):length(l), width(w)
    {cout<<"constructing a Rectangle object..\n";}
    Rectangle& operator = (Rectangle& r);
    int getLength() const { return length;}
    int getWidth() const { return width; }
    void setLength(int l) { length = l; }
    void setWidth(int w) { width = w; }
    friend Rectangle operator + (Rectangle& r1, Rectangle& r2);
};
Rectangle& operator ++ (Rectangle& r);

int main()
{
    Rectangle rect1(5,6), rect2(7,8), rect3;
    rect3 = rect1 + rect2;
    cout<<"length="<<rect3.getLength()<<" width="<<rect3.getWidth()<<endl;

    ++rect3;
    cout<<"length="<<rect3.getLength()<<" width="<<rect3.getWidth()<<endl;
    return 0;
}
```

Define overloading operators =, + and ++ and write out the output of the program.

3. Write a program that defines a *Date* class to implement the following operations:

(1) Print a date;

(2) Increment the date by one month;

(3) Compare two dates by relation operators ==, <=, and !=;

(4) Overload insertion (>>) and extraction (<<) operators for easy input and output.

(The *Date* class requires the user to input/output a date in the form: dd/mm/yyyy)

4. Given the following program:

```
#include <iostream>
using namespace std;
class CVector {
    public:
        int x,y;
        CVector ();
        CVector (int,int);
        CVector(CVector&);
        CVector& operator ++ ();
        friend CVector operator + (CVector& v1, CVector& v2);
        ~CVector();
};
```

Write out the definitions of all member functions and a friend function within the *CVector* class and test them in the *main* function.

5. Given the declaration of the *Integer* class:

```
class Integer{
public:
        Integer(int a = 0);
        Integer(const Integer& i);
        Integer& operator++ ();
        friend bool operator< (Integer& i, Integer& j);
        void print() const ;
private:
        int value;
};
```

(1) Write the definition of all member functions outside the *Integer* class.

(2) Write the *main* function to test them.

6. Define the *Counter* class including overloading operator ++.

7. When the C++ program is running, it will not check whether an array is out of bound automatically. Write a program that can do it by overloading operator [].

8. Modify the *TimeDemo* class to implement the input and output of the time. Overloading operator >> to input time from the user. Overloading operator << to output the time to console.

9. Design a *Percent* class that represents the percentage of 10%, 80% and etc. Overload operators << and >> to implement the output and input of *Percent* objects; Overload operators = = and < to compare the relationship of two *Percent* objects.

Chapter 6   Inheritance

Do not multiply objects without necessity.
—*W. Occam*

Objectives
- To create classes by inheriting from existing classes
- The use of constructors and destructors in inheritance hierarchies
- Access control in the class and the differences among public, protected and private inheritances
- To understand multiple inheritances
- To avoid ambiguities with multiple inheritances
- To avoid ambiguities with virtual inheritances

6.1   Class Hierarchies

In the previous chapters, we discuss how to build a single class that represents real objects. The class is represented by a collection of abstracted concepts with common characteristics. However, a concept does not exist in isolation always. It coexists with related concepts and derives much of its power from relationships with related concepts. It is a fundamental aspect of human intelligence to seek out, recognize, and create relationships among concepts.

For example, the concepts of a circle, a triangle and a rectangle are related in that they are all shapes; that is, they have the concept of shape in common, that is, they have the operations of the move, draw and calculation. However, we find their different points of the circle, triangle and rectangle. Thus, we must explicitly define classes *Circle*, *Triangle* and *Rectangle* to have class *Shape* in common. Representing a circle, a triangle and a rectangle in a program without involving the notion of shape would be to lose something essential. Each kind of shape is more specialized than its parent *Shape*? We can classify a different kind of shape according to the hierarchy shown in Figure 6-1. These could be represented in the world of classes with a *Shape* class from which we would derive the three other ones: *Circle*, *Triangle* and *Rectangle*.

Class *Shape* contains the members that are common for the three types of shape. *Circle Triangle* and *Rectangle* would be its derived classes, with specific features that are different from one type of polygon to the others.

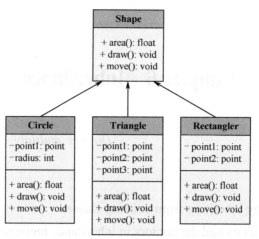

Figure 6-1   A class hierarchy

Such a set of related classes is traditionally called a ***class hierarchy***. A class hierarchy is most often a tree (Figure 6-2 (a)), but it can also be a more general graph structure (Figure 6-2 (b)).

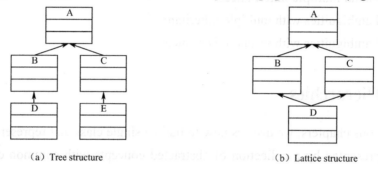

(a) Tree structure (b) Lattice structure

Figure 6-2   The structure of class hierarchies

A class hierarchy is created by inheritance. *Inheritance* is a key feature of C++ classes. It can allow programmers to reuse code they've already written, that is, allow to create classes which are derived from existing classes, so that they automatically include some of its "parent's" members, plus its own members.

Therefore, inheritance suggests an object can inherit characteristics from another object. In more concrete terms, an object can pass on its states and behaviors to its children. For inheritance to work, the objects need to have characteristics in common with each other.

6.2   Derived Classes

6.2.1   Declaration of Derived Classes

In order to derive a class from another, we use a colon (:) in the declaration of the derived class with the following format:

Syntax
```
class derivedClassName : public baseClassName {
    //members;
};
```

where *derivedClassName* is the name of the derived class (new class) and *baseClassName* is the name of the base class (existing class). The *public* access specifier may be replaced by any one of the other access specifiers *protected* and *private*. This access specifier limits the most accessible level for the members inherited from the base class. The members with a more accessible level are inherited with this level instead, while the members with an equal or more restrictive access level keep their restrictive level in the derived class.

6.2.2 Structure of Derived Classes

For example, we consider building a program that deals with an ordinary person. By using the data abstraction, we can define a *Person* class. Such a program might have a class like this:

```
class Person{
private:
    string first Name;
    string familyName;
    int age;
    //...
};
```

We can declare a *Student* class. We consider the *Student* has the properties of *Person*. Thus, we might try to define a student as follows:

```
class Student{
private:
    Person person;   //student's person record
    String schoolName;
    //...
};
```

As a student is also a person, the *Person* class data is stored as the person member of a *Student* object. This may be obvious to a human reader—especially a careful reader—but there is nothing that tells the compiler and other tools that a student is also a person. A *Student* object is not a *Person* object, so one cannot simply create one object through the object of another class.

In the real world, we do not view everything as unique; we often view something as being like something else but with differences or additions. Students are like regular persons; however, there might be differences.

The correct approach is to explicitly state that a *Student* is a *Person*, with an inheritance relationship. So, let us redefine the *Student* class:

```
class Student: public Person{
private:
    string schoolName;
};
```

Class *Student* is derived from class *Person*, and conversely, *Person* is a base class for *Student*. Class *Student* has the members of class *Person*, such as *firstName*, *familyName*, etc., in addition to its own member, such as *schoolName*.

A class which adds new functionality to an existing class is said to derive from that original class. The original class is said to be the new class' *base class*. Thus, a derived class is often said to inherit properties from its base, so the relationship is also called **inheritance**. A **base class** is sometimes called a **superclass** and a **derived class** is called a **subclass**.

> **Inheritance** is a mechanism by which one class acquires the properties (the data and operations) of an existing class.
> **Base class** (superclass)—the class being inherited from.
> **Derived class** (subclass)—the class that inherits.
> 继承是一个类从已有类获得属性（数据和操作）的机制。
> 基类（父类）——被继承的类。
> 派生类（子类）——继承的类。

This terminology, however, is confusing to people who observe that the data in a derived class object is a superset of the data in its base class object. A derived class is larger than its base class in the sense that it holds more data and provides more functions.

A popular and efficient implementation of the notion of derived classes has an object of the derived class represented as an object of the base class, with the information belonging specifically to the derived class added at the end, as shown in Figure 6-3.

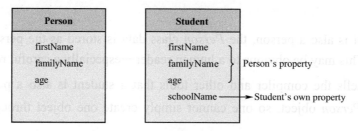

Figure 6-3 Classes *Person* and *Student*

Example 6-1: Deriving a *Student* class from a *Person* class.

```
//--------------------------------------------------------------
//File: Example6_1.cpp
//This program defines a Person class and a Student class.
```

```cpp
//----------------------------------------------------------------------
1   #include<iostream>
2   #include<string>
3   using namespace std;
4
5   class Person{
6   private:
7       string firstName;
8       string familyName;
9       int age;
10  public:
11      //declaration of member functions
12  };
13  class Student: public Person{
14  private:
15      string schoolName;
16  public:
17      //declaration of member functions
18  };
19  void g(Student st, Person per)
20  {
21      Person pe = st;         //ok: every Student is a Person
22      //Student pm = per;     //error: not every Person is a Student
23  }
24  int main()
25  {
26      Person Tom;
27      Student Jack;
28      g(Jack, Tom);
29      return 0;
30  }
```

Example 6-1 has no output because we haven't written any functions. The statements in Lines 5 through 12 define a *Person* class, and those in Lines 13 through 18 define a *Student* class. Line 13 defines a *Student* class that derived from the *Person* class.

Deriving class *Student* from class *Person* in this way makes class *Student* be a subtype of class *Person* so that a *Student* class can be used wherever a *Person* class is acceptable. A *Student* is (also) a *Person*, so a *Student* can be used as a *Person*. However, a *Person* class is not necessarily a *Student* class, so a *Person* cannot be used as a *Student*.

In general, if a derived class has a public base class, then its object (e.g. a *Student* object) can be assigned to a variable of the base class (a *Person* object) without the use of explicit type conversion, like the statement in Line 21. The opposite conversion, from *Person* to *Student*, the object of the base class cannot be assigned to the derived class object, like the statement in Line 22.

Notice that an object of a derived class can be treated through pointers and references. The opposite is not true. For example,

```
void g (Student st, Person per)
{
    Person* pe = &st;        //ok: every Student is a Person
    Student* pm = &per;      //error: not every Person is a Student
}
```

Using a base class is equivalent to declaring an (unnamed) object of that class. Consequently, a class must be defined in order to be used as a base:

```
class Person;                    //declaration only, no definition
class Student: public Person{    //error: Person not defined
//...
};
```

6.3 Constructors and Destructors of Derived Classes

6.3.1 Constructors of Derived Classes

In principle, a derived class inherits every member of its base class except:
- its constructor and its destructor
- its operator= members
- its friends

Derived Classes with a Default Constructor

A constructor plays a vital role in initializing an object. Some derived classes need constructors. If a base class has constructors, then they must be invoked. Although the constructors and destructors of the base class are not inherited themselves, its default constructor (i.e., its constructor without parameters) and its destructor are always called when a new object of a derived class is created or destroyed.

The constructor of a derived class is defined in the following way:

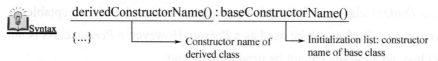

Example 6-2: Definition of a derived class with a default constructor.

```
//----------------------------------------------------------------
//File: Example6_2.cpp
//This program defines a Person class and a Student class with default constructors.
//----------------------------------------------------------------
1    #include <iostream>
```

```
2     using namespace std;
3
4     class Person{
5     public:
6         Person()            //default constructor
7         {
8             cout << "Constructor of the base class!" << endl;
9         }
10    };
11    class Student:public Person{
12    public:
13        Student():Person()    //initialize base class
14        {
15            cout << "Constructor of the derived class!" << endl;
16        }
17    };
18
19    int main()
20    {
21        Person Tom;
22        Student Jack;
23        return 0;
24    }
```

Result:

```
1    Construction of the base class!
2    Construction of the base class!
3    Construction of the derived class!
```

Notice that the constructor of *Person* in Line 13 can be omitted because the constructor of the base class has no parameters. Thus, the statement in Line 13 can be written in the following form:

```
Student();
```

Derived Classes with Constructors

However, if all constructors for a base require parameters, then a constructor for that base must be explicitly called, and you can specify it in each constructor definition of the derived class.

In this case, the syntax of the constructor definition of the derived class is

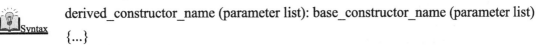

derived_constructor_name (parameter list): base_constructor_name (parameter list)
{...}

Parameters of the base class' constructor are specified in the definition of a derived class' constructor. In this respect, the base class acts exactly like a member of the derived class. For example, the constructor of the *Person* base class is defined as follows:

Person::Person (const string&n, int a) :**familyName (n), age (a) {}**
                                         ↓
                                    initialization list

The constructor of the *Student* derived class is defined as follows:

```
Student::Student(const string& n, int a, const string &sn): Person (n, a)    //initialize base
{
    schoolName = sn;        //initialize the members of the derived class
}
```

If the constructor of the derived class directly initializes the member of its base class like the following way,

```
Student: Student (const string& n, int a, const string &sn)
{
    familyName = n;        //error: familyName not declared in Student
    age = a;               //error: age not declared in Student
    schoolName = sn;       //ok
}
```

Then this definition contains three errors: it fails to invoke *Person*'s constructor, and twice it attempts to initialize the members of *Person* directly.

A derived class constructor can specify the initializer for its own members and immediate bases only; it cannot directly initialize the members of its base class.
派生类构造函数只能为自己的成员和它的基类指定初始化方式；它不能直接初始化其基类的成员。

Example 6-3: Definition of the derived class constructor with parameters.

```
//-------------------------------------------------------------------
//File: Example6_3.cpp
//This program defines the constructors of a Person class and a Student class.
//-------------------------------------------------------------------
1    #include <iostream>
2    using namespace std;
3
4    class Person{
5    private:
6        string familyName;
7        int age;
8    public:
9        Person(const string& n, int a)
10       {
11           familyName = n;        //initialize the members
12           age = a;
```

```
13      }
14    };
15    class Student: public Person{
16    private:
17        string schoolName;
18    public:
19        Student(const string& n, int a, const string &sn):Person(n, a)    //initialize base class
20        {
21            schoolName = sn;          //initialize the data member
22        }
23    };
24
25    int main()
26    {
27        Person Tom("Tom",10);
28        Student Jack("Jack",10, "SUT");
29        return 0;
30    }
```

6.3.2 Destructors of Derived Classes

When a derived class object is destroyed, the program calls that object's destructor. This begins a chain (or cascade) of destructor calls in which the derived class destructor, the destructors of the direct and indirect base classes, and the classes' members execute in the reverse order in which the constructors executed. When a derived class object's destructor is called, the destructor performs its task, and then invokes the destructor of the next base class up the hierarchy. This process repeats until the destructor of the final base class at the top of the hierarchy is called. Then the object is removed from memory.

Example 6-4: Definition of derived class constructor and destructor.

```
//---------------------------------------------------------------------------
//File: Example6_4.cpp
//This program defines a Person class and a Student class.
//---------------------------------------------------------------------------
1    #include<iostream>
2    using namespace std;
3
4    class Person{
5        string firstName;
6        int age;
7    public:
8        Person(const string& n, int a)
9        {
10            firstName = n;          //initialize members
11            age = a;
```

```
12      }
13      ~Person()
14      {
15          cout << "Destructor of the Person class " << firstName << endl;
16      }
17  };
18  class Student : public Person{
19      string schoolName;
20  public:
21      Student(const string& n, int a, const string &sn) : Person(n, a)
22      {
23          schoolName = sn;            //initialize members
24      }
25      ~Student()
26      {
27          cout << "Destructor of the Student class" << endl;
28      }
29  };
30
31  int main()
32  {
33      Person Tom("Tom",10);
34      Student Jack("Jack",10, "NEU");
35      Student Jenny("Jenny",10, "SUT");
36      return 0;
37  }
```

Result:

```
1  Destructor of the Student class
2  Destructor of the Person class Jenny
3  Destructor of the Student class
4  Destructor of the Person class Jack
5  Destructor of the Person class Tom
```

6.3.3 The Calling Order of Derived Class Objects

The derived class objects are constructed from the bottom up: first the base class, then the members, and then the derived class itself. They are destroyed in the opposite order: first the derived class itself, then the members, and then the base class. Members and the base class are constructed in order of declaration in the class and destroyed in the reverse order.

Let us look at the following example about the calling order in a class hierarchy.

Example 6-5: Calling order of class objects.

```
//------------------------------------------------------------------------
//File: example6_5.cpp
```

//This program defines a *Person* class and a *Student* class.
//--

```
1    #include <iostream>
2    #include <string>
3    using namespace std;
4    class Date{
5        int day;
6        int month;
7        int year;
8    public:
9        Date(int d=1,int m=1,int y=2010) : day(d), month(m), year(y)
10       {
11           cout << "Date's constructor is called" << endl;
12       }
13       ~Date()
14       {
15           cout << "Date's destructor is called" << endl;
16       }
17   };
18   class Person{
19       string name;
20   public:
21       Person() : name("noname")
23       {
24           cout << "Person's default constructor: Person's name is " << name << endl;
25       }
26       Person(const string &na) : name(na)
27       {
28           cout <<"Person's constructor: Person's name is " << name << endl;
29       }
30       ~Person()
31       {
32           cout << "Person's destructor: Person's name is " << name << endl;
33       }
34   };
35   class Student : public Person{
36       Date enrolldate;           //Date object
37       string schoolName;
38   public:
39       Student()
40       {
41           cout << "Student's default constructor is called" << endl;
42       }
43       Student(const string &na,int d,int m,int y,const string &sn) :
44           Person(na), enrolldate(d,m,y), schoolName(sn)
45       {
46           cout << "Student's constructor is called" << endl;
```

```
47          }
48          ~Student()
49          {
50              cout << "Student's destructor is called" << endl;
51          }
52      };
53      int main()
54      {
55          Student noname;
56          Student Jack("Jack", 12, 12, 1991, "SUT");
57          return 0;
58      }
```

Result:

```
1   Person's default constructor:Person's name is noname
2   Date's constructor is called
3   Student's default constructor is called
4   Person's constructor:Person's name is Jack
5   Date's constructor is called
6   Student's constructor is called
7   Student's destructor is called
8   Date's destructor is called
9   Person's destructor:Person's name is Jack
10  Student's destructor is called
11  Date's destructor is called
12  Person's destructor:Person's name is noname
```

In Line 55, object *noname* of the *Student* class is created, printing out the first three lines of output when the *Student*'s default constructor is called. Line 39 states the *Student*'s default constructor. The base's constructor can be omitted in the Student's initializers because *Person* also has a default constructor. When we call the *Student*'s default constructor, the *Person*'s default constructor will be called implicitly although we do not call it explicitly, and then the *Date*'s default constructor will be called implicitly.

In Line 56, object *Jack* is created, printing out the lines 4 through 6 of results when the *Student*'s constructor is called. Line 43 states the *Student*'s constructor. When we called the *Student*'s constructor, the *Person*'s constructor will be called first, and then the *Date*'s constructor will be called.

Finally, the *Student* object goes out of scope and the destructor is called. The objects are destroyed in the opposite order. Firstly, the *Student*'s destructor is called. Secondly, the *Date*'s destructor is called. Last, the *Person*'s destructor is called.

In a word, the construction order of the *Student* object is as follows: *Person→Date→Student*. Whereas, the destruction order of the *Student* object is as follows: *Student→Date→Person*.

6.3.4 Inheritance and Composition

It turns out that much of the syntax and behavior are similar for both composition and inheritance (which makes sense as they are both ways of making new classes from existing classes). However, composition and inheritance are different.

Inheritance represents the *is a* relationship. In the *is a* relationship, an object of a derived class also can be treated as an object of its base class. For example, a student is a person, so any properties and behaviors of a person are also the properties of a student.

By contrast, ***composition*** (a kind of aggregation) represents a "***has a***" relationship. In a "*has a*" relationship, an object contains one or more objects of other classes as members. In Example 6-5, a *Student* class has an *enrolldate* property (in Line 36) which is an object of class *Date*. In other words, we can see the *enrolldate* object is a data member of the *Student* class.

The UML diagram of these two kinds of class relationship is shown in Figure 6-4.

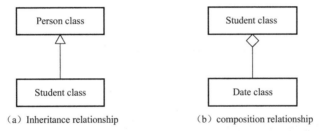

(a) Inheritance relationship   (b) composition relationship

Figure 6-4 The UML diagram of two kinds of class relationship

---

Think These Over

1. If a base class has a default constructor, does the constructor in the derived class need to explicitly call a base class constructor? If a base class has a constructor with parameters but no default values, does the derived class need to call a constructor with parameters explicitly?

2. How do we differentiate between composition and inheritance?

3. What do the advantages of inheritance and composition include?

6.4 Member Functions of Derived Classes

6.4.1 Defining a Member Function

We need to define the member functions of classes *Person* and *Student*. For example,

```
class Person{
private:
```

```
            string firstName;
            string familyName;
public:
    void print() const;
    string full_name() const
    { return firstName + familyName; }
};
class Student: public Person{
private:
    string schoolName;
public:
    void print() const;
};
```

In this example, the *print* function of derived class *Student* can use the public or protected members of its base *Person* in the derived class itself. For example,

```
void Student::print() const
{
    cout << "name is " << full_name() << endl;
}
```

However, the *print* function of derived class *Student* cannot use a base class' private members.

```
void    Student::print() const
{
    cout << "name  is" << familyName << endl;        //error
}
```

This second version of *Student::print()* would no compile. A member of a derived class has no special permission to access private members of its base class, so *familyName* is not accessible to *Student::print()*.

Typically, the cleanest solution is that the derived class uses only the public members of its base class. For example,

```
void Student::print() const
{
    Person::print();              //print Person information
    cout << schoolName;           //print Student-specific information
}
```

The **scope operator (::)** must be used because function *print* has been redefined in the derived class *Student*. Such reuse of names is typical.

作用域运算符(::)必须被使用，因为 *print* 函数在 *Student* 派生类中被重新定义了。这样的名字重用是常见的。

· 176 ·

6.4.2　Overriding Member Functions

A *Student* object has access to all member functions in the *Person* class, as well as to any member functions, such as *full_name*, which is the function declaration that the *Student* class might add.

Sometimes, to keep the same function name as that of its base class, the *Student* class can also **override** a base class function. ***Overriding a member function*** means changing the implementation of a base class function in a derived class. When you make an object of the derived class, the correct function is called.

When you override a function, it must agree in returning-value type and in signature with the function in the base class. The ***signature*** is the function prototype rather than the returning-value type: that is, the name, the parameter list, and the ***const*** keyword if used.

> When a derived class creates a function with the same returning-value type and signature as a member function in the base class, but with a new implementation, it is said to be **overriding** that method. The **signature** of a function is its name, as well as the number and type of its parameters. The signature does not include the returning-value type.
>
> 当派生类要创建一个与基类成员函数具有相同返回类型和签名的成员函数时，但该函数提供新的实现内容，该方法被称为**覆写**（overriding）。函数的**签名**（signature）由函数名字、参数个数和参数类型组成，签名不包含返回类型。

Example 6-6: Overriding the *print* function of a *Student* class.

```
//---------------------------------------------------------------
//File: Example6_6.cpp
//This program defines a Person class and a Student class.
//---------------------------------------------------------------
1   #include<iostream>
2   include <string>
3   using namespace std;
4
5   class Person {
6       string firstName;
7       string familyName;
8       int age;
9   public:
10      Person(const string &firstname, const string &familyname, int a)
11      {
12          firstName = firstname;
13          familyName = familyname;
14          age = a;
15      }
16      void print() const
```

· 177 ·

```
17    {
18        cout << "name is " << full_name() << endl;
19        cout << "age is " << age << endl;
20    }
21    string full_name() const
22    {   return firstName + ' ' + familyName;   }
23 };
24 class Student: public Person {
25    string schoolName;
26 public:
27    Student(const string &firstname, const string &familyname, int age, const string &sn) :
28        Person(firstname, familyname, age)
29    {
30        schoolName = sn;
31    }
32    void print() const     //overriding the print function
33    {
34        Person::print();
35        cout << schoolName << endl;
36    }
37 };
38 int main()
39 {
40    Person Tom("Tom", "Felton", 20);
41    Student Jack("Jack", "Sparrow", 15, "SUT");
42    Tom.print();
43    Jack.print();
44    return 0;
45 }
```

Result:

1	name is Tom Felton
2	age is 20
3	name is Jack Sparrow
4	age is 15
5	SUT

In Line 32, derived class *Student* overrides the *print* function. In Line 40, when a *Person* object *Tom* is created, the *Person* constructor is called. In Line 41, when a *Student* object *Jack* is created, the *Person* constructor and then the *Student* constructor are called respectively.

In Line 42, the *Person* object calls its *print* function, and the *Student* object then calls its *print* function in Line 43. The output reflects that the correct functions were called. Finally, the two objects go out of scope and their destructors are called respectively.

If you have overridden the base function, it is still possible to call it by fully qualifying the name of the method. You do this by writing the base class' name, followed by two colons and then the method name. For example, *Person::print()*.

Overriding Functions and Overloading Functions

When you override a function of the derived class, the corresponding functions must have the same return type, name and parameter list.

When you overload a function of the derived class, the function can have the same name and different parameter list.

Think These Over

Why do we need to override the member functions of the derived class?

6.5 Access Control

6.5.1 Access Control in Classes

As mentioned in §3.4, a member of a class can be ***private***, ***protected***, or ***public***:

- If it is ***private***, its name can be used only by member functions and friends of the class in which it is declared.
- If it is ***protected***, its name can be used only by member functions and friends of the class in which it is declared and be accessed by member functions and friends of classes derived from this class.
- If it is ***public***, its name can be used by any function.

This reflects the view that there are three kinds of functions accessing a class: functions implementing the class (its friends and members), functions implementing a derived class (its friends and members), and the other functions. This can be presented graphically, shown in Figure 6-5.

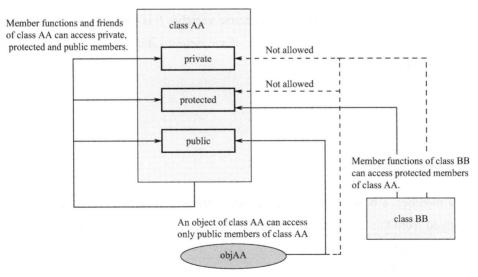

Figure 6-5   Three kinds of access control of a class

Example 6-7: Access control to the members of a *samp* class.

```
//------------------------------------------------------------------
//File: example6_7.cpp
//This program verifies access control to the members of a samp class.
//------------------------------------------------------------------
1    #include <iostream>
2    using namespace std;
3
4    class samp{
5        int a;
6    protected:
7        int b;
8    public:
9        int c;
10       samp(int n,int m)
11       {   a = n;   b = m;   }
12       int geta() {return a; }
13       int getb() {return b; }
14   };
15   int main()
16   {
17       samp sa(20, 30);
18       sa.b=99;                              //error
19       sa.c=50;
20       cout << sa.geta() << endl;
21       cout << sa.getb() << endl;
22       cout << sa.c << endl;
23       return 0;
24   }
```

This example would not be compiled because variable *b* is the protected member of class *samp*. It cannot be accessed outside the class. If we delete Line 18, we will see the output of the example like this:

```
1    20
2    30
3    50
```

6.5.2 Access to Base Classes

Like a member, a base class can be declared ***private***, ***protected***, or ***public*** when another class is derived from this base class. For example,

```
class X : public B {/*        */};
class Y : protected B {/*     */};
class Z : private B {/*       */};
```

The keyword after the colon (:) denotes the most accessible levels of the members inherited from these class.

Public derivation makes the derived class a subtype of its base; this is the most common form of derivation. Since ***public*** is the most accessible level, by specifying this keyword the derived class will inherit all the members with the same access levels they had in the base class.

Protected and ***private derivations*** are used to represent implementation details. ***Protected*** bases are useful when the further derivation is needed in the class hierarchies. ***Private*** bases are most useful when defining a class by restricting the interface to a base so that stronger guarantees can be provided.

The access specifier for a base class can be left out. In that case, the base defaults to a ***private*** base for a class and a public base to a struct.

For example,

```
class XX : B {/*...*/};     //B is a private base for a class type
struct YY : B {/*...*/};    //B is a public base for a struct type
```

For readability, it is best always to use an explicit access specifier.

For example,

```
class base : private B{};       //B is a private base
class base : protected B {};    //B is a protected base
class base : public B{};        //B is a public base
```

If a member is private in the base class, then only the base class and its friends may access that member. The derived class has no access to the private members of its base class, nor can it make those members access to its own users.

If a base class member is public or protected, then the access label used in the derivation list determines the access level of that member in the derived class:

• In public inheritance, the members of the base class retain their access levels: The public members of the base class are public members of the derived and the protected members of the base are protected in the derived.

• In protected inheritance, the public and protected members of the base class are protected members in the derived class.

• In private inheritance, all the members of the base class are private in the derived class.

We can see this clearly in Table 6-1.

Table 6-1 Access control

	Base class	Derived class		General users
Private Inheritance	Private members	Private members	NA	NA
	Protected members	Private members	A	NA
	Public members	Private members	A	NA
Protected Inheritance	Private members	Private members	NA	NA
	Protected members	Protected members	A	NA
	Public members	Protected members	A	NA
Public Inheritance	Private members	Private members	NA	NA
	Protected members	Protected members	A	NA
	Public members	Public members	A	A

A—Access   NA—No access

Example 6-8: Three kinds of inheritance.

```
//-------------------------------------------------------------------------------
//File: Example6_8.cpp
//This program defines some classes to test classes in the way to different inheritance.
//-------------------------------------------------------------------------------
1    class Base{
2    public:
3       int m1;
4    protected:
5       int m2;
6    private:
7       int m3;
8    };
9    class Privateclass : private Base{
10   public:
11      void test()
12      {
13         m1 = 1;          //ok
14         m2 = 2;          //ok
15         m3 = 3;          //error
16      }
17   };
18   class DerivedFromPri : public Privateclass{
19   public:
20      void test()
21      {
22         m1 = 1;          //error
23         m2 = 2;          //error
24         m3 = 3;          //error
25      }
```

```cpp
26      };
27      class Protectedclass : protected Base{
28      public:
29          void test()
30          {
31              m1 = 1;         //ok
32              m2 = 2;         //ok
33              m3 = 3;         //error
34          }
35      };
36      class DerivedFromPro : public Protectedclass{
37      public:
38          void test()
39          {
40              m1 = 1;         //ok
41              m2 = 2;         //ok
42              m3 = 3;         //error
43          }
44      };
45      class Publicclass : public Base{
46      public:
47          void test()
48          {
49              m1 = 1;         //ok
50              m2 = 2;         //ok
51              m3 = 3;         //error
52          }
53      };
54      class DerivedFromPub : public Publicclass{
55      public:
56          void test()
57          {
58              m1 = 1;         //ok
59              m2 = 2;         //ok
60              m3 = 3;         //error
61          }
62      };
63      int main()
64      {
65          Privateclass pri;
66          pri.m1 = 1;         //error
67          pri.m2 = 2;         //error
68          pri.m3 = 3;         //error
69          Protectedclass pro;
70          pro.m1 = 1;         //error
71          pro.m2 = 2;         //error
72          pro.m3 = 3;         //error
```

```
73      Publicclass pub;
74      pub.m1 = 1;
75      pub.m2 = 2;        //error
76      pub.m3 = 3;        //error
77      return 0;
78    }
```

This example will not compile because the members of part of the classes above are not accessed by the *main* function. Please try to correct and give its result. The levels of inheritance in example 6-8 are shown in Figure 6-6.

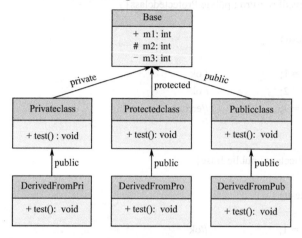

Figure 6-6 Levels of inheritance

Think It Over

What is the effect of an access specifier?

6.6 Multiple Inheritance

In the previous sections, each class has inherited from a single parent. Such single inheritance is enough to describe most real-world relationships. Some classes, however, represent the blending of two classes into one.

Sometimes a class is constructed from a lattice of base classes. A derived class itself can be a base class. For example,

```
class Employee{/*...*/};
class Manager: public Employee{/*...*/};
class Director: public Manager{/*...*/};
```

Because most such lattices historically have been trees, a class lattice is often called a *class hierarchy*. It can also be a more general graph structure, as shown in Figure 6-7. For example,

```
class Temporary {/*...*/};
class Employee {/*...*/};
class Secretary : public Employee {/*...*/};
class Manager : public Employee {/*...*/};
class Tsec : public Temporary, public Secretary {/*...*/};
class Consultant : public Temporary, public Manager {/*...*/};
class Director : public Manager {/*...*/};
```

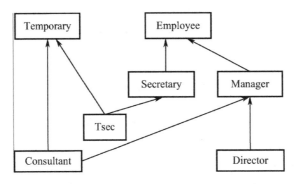

Figure 6-7　A class hierarchy

6.6.1　Declaration of Multiple Inheritance

In C++, it is perfectly possible that a class inherits members from more than one class. This is done by simply separating the different base classes with commas in the derived class declaration.

Multiple inheritance is defined in the following way:

   class derived_class: public base_class1, public base_class2
   { /*...*/ };

where *base_class1* and *base_class2* are the base classes' name. They are separated by commas. Access specifiers precede each base class.

For example, a *SleeperSofa* class is considered as follow:

```
class SleeperSofa : public Bed, public Sofa{ };
```

As the name implies, it is a sofa and a bed (although not a very comfortable bed). Thus, the sleeper sofa should be allowed to inherit the bed like properties.

> C++ allows a derived class to inherit from more than one base class. This is called **multiple inheritance.**
> C++允许派生类从多个基类继承。这就是**多继承**。

To see how multiple inheritance works, look at the sleeper sofa example. Figure 6-8 shows the inheritance graph for class *SleeperSofa*. Notice how this class inherits from classes *Sofa* and *Bed*. In this way, it inherits the properties of both.

The code to implement class *SleeperSofa* looks like the following example.

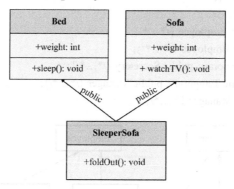

Figure 6-8   Class hierarchy of class *SleeperSofa*

Example 6-9: Definition of multiple-inherited class *SleeperSofa*.

```
//---------------------------------------------------------------------------
//File: example6_9.cpp
//This program defines two base classes Bed and Sofa and a multiple-inherited class SleeperSofa.
//---------------------------------------------------------------------------
1   #include <iostream>
2   using namespace std;
3   class Bed{
4   public:
5       Bed(){}
6       void sleep(){ cout << "Sleep" << endl; }
7       int weight;
8   };
9   class Sofa{
10  public:
11      Sofa(){}
12      void watchTV(){ cout << "Watch TV" << endl; }
13      int weight;
14  };
15  class SleeperSofa : public Bed, public Sofa{
16  public:
17      SleeperSofa(){}
18      void foldOut(){ cout << "Fold out" << endl; }
19  };
20  int main()
21  {
22      SleeperSofa ss;
23      //you can watch TV on a sleeper sofa like a sofa...
24      ss.watchTV();              //Sofa::watchTV()
25      //you can fold it out...
```

26	ss.foldOut();	//SleeperSofa::foldOut()
27	//sleep on it	
28	ss.sleep();	
29	return 0;	
30	}	

Class *SleeperSofa* inherits from both *Bed* and *Sofa*. This is apparent from the appearance of both classes in the class declaration. *SleeperSofa* inherits all the members of both base classes. Thus, both of the calls *ss.sleep* and *ss.watchTV* are legal. You can use a *SleeperSofa* as a *Bed* or a *Sofa*. Additionally, class *SleeperSofa* can have its own member, such as *foldOut*. The output of this program appears as follows:

1	Watch TV
2	Fold out
3	Sleep

6.6.2 Constructors of Multiple Inheritance

The constructing order of the multiple inheritance is similar to the single inheritance.
The syntax of the constructor of a derived class is

Syntax

   constructor (parameter): base_class1's constructor(parameter),
         base_class2's constructor(parameter),
         base_class3's constructor(parameter)

 { /*...*/ }

Example 6-10: The constructor of multiple-inherited class *SleeperSofa*.

```
//-----------------------------------------------------------------
//File: example6_10.cpp
//This program defines two base classes Bed and Sofa and a multiple-inherited class SleeperSofa.
//-----------------------------------------------------------------
1   #include <iostream>
2   using namespace std;
3
4   class Bed{
5   public:
6       Bed(int w): weight(w) { cout << "Constructing Bed object...\n"; }
7       void sleep()
8       { cout << "Sleep...\n"; }
9       int weight;
10  };
11  class Sofa{
12  public:
13      Sofa(int w) : weight(w) { cout << "Constructing Sofa object...\n"; }
14      void watchTV()
15      { cout << "Watching TV...\n"; }
```

```
16      int weight;
17    };
18    class SleeperSofa : public Bed, public Sofa{
19    public:
20       SleeperSofa(int w) : Bed(w), Sofa(w)
21       { cout << "Constructing SleeperSofa object...\n"; }
22       void foldOut()
23       { cout << "Fold out...\n"; }
24    };
25    int main()
26    {
27       SleeperSofa ss(2);
28       ss.sleep();
29       ss.watchTV();
30       ss.foldOut();
31       return 0;
32    };
```

Result:

```
1    Constructing Bed object...
2    Constructing Sofa object...
3    Constructing SleeperSofa object...
4    Sleep...
5    Watching TV...
6    Fold out...
```

From the observation of Example 6-10, when a class has multiple base classes, base classes are initialized in a depth-first, left-to-right order of appearance in the base-specifier-list. In Example 6-10, when the *ss* object is created, the order of construction is as follows: *Bed* → *Sofa* → *SleeperSofa*.

6.7 Virtual Inheritance

6.7.1 Multiple Inheritance Ambiguities

Although multiple inheritance is a powerful feature, it introduces several possible problems. One is apparent in the preceding example. Notice that both *Bed* and *Sofa* contain a member *weight* in Example 6-10. This is logical because both have a measurable weight. The question is, "which *weight* does *SleeperSofa* inherit?" The answer is "both". *SleeperSofa* inherits a member *Bed::weight* and a separate member *Sofa::weight*. Because they have the same name, unqualified references to *weight* are now ambiguous. This is demonstrated in the following code snippet:

```
1   #include <iostream>
2   void fn()
3   {
4       SleeperSofa ss;
5       cout << "weight = "<< ss.weight << endl;      //illegal - which weight?
6   }
```

The program must now indicate one of the two weights by specifying the desired base class. The following code snippet is correct:

```
1   #include <iostream>
2   void fn()
3   {
4       SleeperSofa ss;
5       cout << "sofa weight = " << ss.Sofa::weight   //specify which weight
6   }
```

Although this solution corrects the problem, specifying the base class in the application function isn't desirable because it forces class information to leak outside the class into application code. In this case, the *fn* function must know that *SleeperSofa* inherits from *Sofa*. These types of so-called **name collisions** weren't possible with single inheritance but are a constant danger with multiple inheritance.

6.7.2 Trying to Solve Inheritance Ambiguities

In the previous case of *SleeperSofa*, the name collision on weight was more than a mere accident. A *SleeperSofa* class doesn't have a bed weight separate from its sofa weight. The collision occurred because this class hierarchy does not completely describe the real-world. Specifically, the classes have not been completely factored.

Thinking about it a little more, it becomes clear that both beds and sofas are special cases of a more fundamental concept: furniture. The *weight* is a property of all furniture. A latticed class hierarchy is described in Figure 6-9.

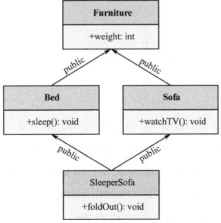

Figure 6-9   Further factoring of *Bed* and *Sofa* (by *weight*)

Factoring out the *Furniture* class should relieve the name collision. A class hierarchy has been generated in the following program.

Example 6-11: Definition of a latticed class hierarchy.

```
//----------------------------------------------------------------
//File: example6_11.cpp
//This program defines four classes Furniture, Bed, Sofa, SleeperSofa.
//----------------------------------------------------------------
1   #include <iostream>
2   using namespace std;
3   //Furniture - more fundamental concept; this class has "weight" as a property
4   class Furniture{
5   public:
6       Furniture(int w) : weight(w) {}
7       int weight;
8   };
9   class Bed : public Furniture{
10  public:
11      Bed(int w) : Furniture(w) {}
12      void sleep(){ cout << "Sleep" << endl; }
13  };
14  class Sofa : public Furniture{
15  public:
16      Sofa(int w) : Furniture(w) {}
17      void watchTV(){ cout << "Watch TV" << endl; }
18  };
19  //SleeperSofa - is both a Bed and a Sofa
20  class SleeperSofa : public Bed, public Sofa{
21  public:
22      SleeperSofa(int w) : Sofa(w), Bed(w) {}
23      void foldOut(){ cout << "Fold out" << endl; }
24  };
25  int main()
26  {
27      SleeperSofa ss(10);
28      //the following is ambiguous; is this a Furniture::Sofa or a Furniture::Bed?
29      cout << "Weight = "<< ss.weight<< endl;      //error
30      return 0;
31  }
```

When a function or data member is called in the shared base class, another ambiguity exists. In Line 29, the *weight* of the *main* function is still ambiguous. If it is not really understood why the *weight* is still ambiguous, try to cast object *ss* of class *SleeperSofa* to a *Furniture* type. The explanation is straightforward. *SleeperSofa* does not inherit from Furniture directly. Both *Bed* and *Sofa* inherit from *Furniture* and then *SleeperSofa* inherits from *Bed* and

Sofa. In memory, a *SleeperSofa* looks like Figure 6-10.

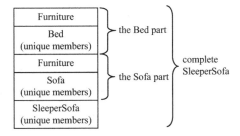

Figure 6-10   Memory layout of a *SleeperSofa*

You can see that a *SleeperSofa* consists of a complete *Bed* followed by a complete *Sofa* followed by some *SleeperSofa* unique stuff. Each of these sub-objects in *SleeperSofa* has its own *Furniture* part, because each inherits from *Furniture*. Thus, a *SleeperSofa* contains two *Furniture* objects!

The hierarchy shown in Figure 6-9 has not been created after all. The inheritance hierarchy created in Example 6-11 is the one as shown in Figure 6-11.

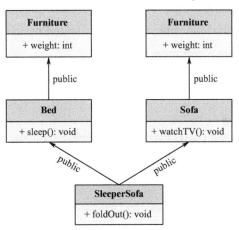

Figure 6-11   Actual result of Example 6-11

6.7.3   Virtual Base Classes

In Example 6-11, the program demonstrates this duplication of the base class. Assume that we write a main function like the following:

```
1   int main()
2   {
3       SleeperSofa ss(10);
4       //solving ambiguous
5       SleeperSofa* pSS = &ss;
6       Sofa* pSofa = (Sofa)*pSS;
7       Furniture* pFurniture = (Furniture)*pSofa;
```

```
        8        cout << "Weight = "<< pFurniture->weight<< endl;
        9        return 0;
       10   }
```

Lines 5 through 7 specifies exactly which *weight* object by recasting the pointer *SleeperSofa* first to a *Sofa** and then to a *Furniture**. However, *SleeperSofa* containing two *Furniture* objects is nonsense. *SleeperSofa* needs only one copy of *Furniture*. We want *SleeperSofa* to inherit only one copy of *Furniture*, and we want *Bed* and *Sofa* to share that one copy. C++ calls this *virtual inheritance* because it uses the *virtual* keyword.

Since a class can be an indirect base class of a derived class more than once, C++ provides a way to optimize the way such base classes work.

> **Virtual base** classes offer a way to save space and avoid ambiguities in class hierarchies that use multiple inheritance. One mechanism to specify a sharing common base is a virtual base class.
>
> 虚基类提供了一种在多继承类层次中可节省空间并避免歧义的方式。虚基类是一种指明共享公共基类的机制。

Armed with this new knowledge, we return to class *SleeperSofa* and implement it as follows. In this example, virtual inheritance is implemented by using a virtual base class, then *Bed* and *Sofa* classes can share a common base.

Example 6-12: Definition of the *SleeperSofa* class with virtual inheritance.

```
//-------------------------------------------------------------
//File: example6_12.cpp
//This program defines four classes Furniture, Bed, Sofa, SleeperSofa in virtual inheritance.
//-------------------------------------------------------------
1    #include <iostream>
2    using namespace std;
3    //Furniture - more fundamental concept; this class has "weight" as a property
4    class Furniture{
5    public:
6        Furniture(int w = 0) : weight(w) {}
7        int weight;
8    };
9    class Bed : virtual public Furniture{
10   public:
11       Bed() {}
12       void sleep(){ cout << "Sleep" << endl; }
13   };
14   class Sofa : virtual public Furniture{
15   public:
16       Sofa(){}
17       void watchTV(){ cout << "Watch TV" << endl; }
18   };
19   //SleeperSofa - is both a Bed and a Sofa
```

```
20      class SleeperSofa : public Bed, public Sofa
21      {
22      public:
23          SleeperSofa(int weight) : Furniture(weight) {}
24          void foldOut(){ cout << "Fold out" << endl; }
25      };
26      int main()
27      {
28          SleeperSofa ss(10);
29          cout << "Weight = " << ss.weight << endl;
30          return 0;
31      }
```

Result:

```
1   Weight=10
```

Notice the addition of the ***virtual*** keyword in class *Bed* and class *Sofa* in Lines 9 and 14 in the inheritance of class *Furniture*. This says, "Give me a copy of *Furniture*, and if you already have one somehow. I will just use that one." A *SleeperSofa* ends up looking like Figure 6-12 in memory.

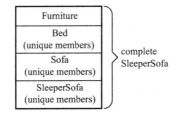

Figure 6-12 Memory layout of *SleeperSofa* with virtual inheritance

Here you can see that *SleeperSofa* inherits *Furniture*, and then *Bed* subtracts the *Furniture* part, followed by *Sofa* minus the *Furniture* part. Bringing up the rear are the members unique to *SleeperSofa*. (Note that this may not be the order of the elements in memory, but that is not important for the purpose of this discussion.)

 DO use multiple inheritance when a new class needs functions and features from more than one base class. DO use virtual inheritance when the most derived classes must have only one instance of the shared base class. DO initialize the shared base class from the most derived classes when using virtual base classes. DON'T use multiple inheritance when single inheritance will finish.

当一个新的类需要继承多个基类的函数和属性时，使用多重继承。当大多数派生类必须只有共享基类的一个实例时，使用虚继承。在使用虚基类时，从最多的派生类开始初始化共享基类。当单个继承能实现时，就不要使用多继承。

6.7.4 Constructing Objects of Multiple Inheritance

The rules for constructing objects need to be expanded to handle multiple inheritance. The constructors are invoked in the following sequence:

First, the constructor for any virtual base classes is called in the order in which the classes are inherited.

Then the constructor for all non-virtual base classes is called in the order in which the classes are inherited.

Next, the constructor for all member objects is called in the order in which the member objects appear in the class.

Finally, the constructor for the class itself is called.

Members and objects are destroyed in the reverse order of the constructor.

Base classes are constructed in the order in which they are inherited but not in the order in which they appear on the constructor line.

基类按照被继承的顺序构造，而不是按照在构造函数行中出现的顺序构造。

Example 6-13: Definition of the *SleeperSofa* class with virtual inheritance.

```
//---------------------------------------------------------
//File: example6_13.cpp
//This program defines defines four classes Furniture, Bed, Sofa, SleeperSofa in virtual inheritance.
//---------------------------------------------------------
1    #include <iostream>
2    using namespace std;
3    class Furniture{
4    public:
5      Furniture(int w) :weight(w)
6      { cout << "Furniture's constructor is called" << endl; }
7      ~Furniture()
8      { cout << "Furniture's destructor is called" << endl; }
9    private:
10     int weight;
11   };
12   class Bed : virtual public Furniture{
13   public:
14     Bed(int w):Furniture(w)
15     { cout << "Bed's constructor is called" <<endl; }
16     ~Bed()
17     { cout << "Bed's destructor is called" << endl; }
18   };
```

· 194 ·

```
19    class Sofa : virtual public Furniture{
20    public:
21       Sofa(int w):Furniture(w)
22       { cout << "Sofa's constructor is called" << endl; }
23       ~Sofa()
24       { cout << "Sofa's destructor is called" << endl; }
25    };
26    //SleeperSofa - is both a Bed and a Sofa
27    class SleeperSofa : public Bed, public Sofa{
28    public:
29       SleeperSofa(int w, bool f):Furniture(w), Bed(w), Sofa(w), fold(f)
30       { cout << "SleeperSofa's constructor is called" << endl; }
31       ~SleeperSofa()
32       { cout << "SleeperSofa's destructor is called" << endl; }
33    private:
34       bool fold;     //if fold? true:fold, false: not fold
35    };
36    int main()
37    {
38       SleeperSofa ss(10, true);
39       return 0;
40    }
```

Result:

1	Furniture's constructor is called
2	Bed's constructor is called
3	Sofa's constructor is called
4	SleeperSofa's constructor is called
5	SleeperSofa's destructor is called
6	Sofa's destructor is called
7	Bed's destructor is called
8	Furniture's destructor is called

Think These Over

1. How do we avoid an inheritance ambiguity in multiple inheritance?

2. In what order is the derived object constructed in multiple inheritance?

6.8 Case Study: The iWatch Class

The Apple Watch is the ultimate device for a healthy life. We design a simple *iWatch* class for the Apple Watch. The *iWatch* class contains your location (i.e. city name), weather and an *ExtTime* class object. The *iWatch* class has the following operations:

(1) Show the current time in China;

(2) Show the time in another zone corresponding to the current time.

Here, the *ExtTime* class is derived from a *Time* class. Their class specification is presented in the UML diagram, as shown in Figure 6-13.

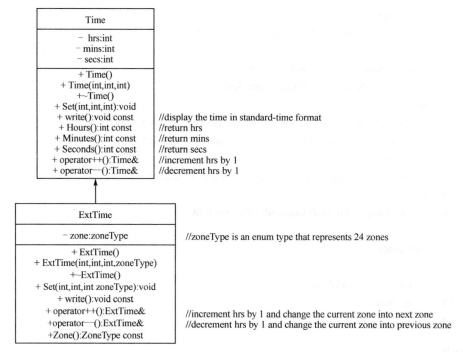

Figure 6-13 The UML diagram of the *Time* and *ExtTime* classes

According to the UML diagram, the *iWatch* class is defined as follows.

Example 6-14: The definition of the *iWatch* class and a test program.

```
//-----------------------------------------------------------------------------
//File: iWatch.h
//This program describes the definition of classes Time, ExtTime and iWatch.
//-----------------------------------------------------------------------------
1  #ifndef _IWATCH_H
2  #define _IWATCH_H
3  #include <iostream>
4  #include <string>
5  using namespace std;
6
7  enum ZoneType { Zone0, EZone1, EZone2, EZone3, EZone4, EZone5, EZone6, EZone7,
8      EZone8, EZone9, EZone10, EZone11, EWZone12, WZone11, WZone10, WZone9,
9      WZone8, WZone7, WZone6, WZone5, WZone4, WZone3, WZone2, WZone1};
10
11 class Time{
12 public:
13     Time();
14     Time(int, int, int);
```

```
15      ~Time();
16      void Set(int, int, int);
17      Time& operator++();
18      Time& operator--();
19      void Write() const;
20
21      int Hours();
22      int Minutes();
23      int Seconds();
24   private:
25      int hrs, mins, secs;
26   };
27
28   class ExtTime : public Time{         //public inheritance
29   public:
30      ExtTime();
31      ExtTime(int, int, int, ZoneType);
32      ~ExtTime();
33      void Set(int, int, int, ZoneType);   //overloading member function
34      ExtTime& operator++();
35      ExtTime& operator--();
36      void Write() const;                  //overriding member function
37      ZoneType Zone();
38   private:
39      ZoneType zone;
40   };
41
42   class iWatch{
43   public:
44      iWatch();
45      ~iWatch();
46      void changeZone(int);
47      void show() const;
48      void showTime() const;
49   private:
50      string cityName;
51      int temperature;
52      ExtTime eTime;                       //object member
53   };
54   #endif
```

//--
//File: **iWatch.cpp**
//The program presents the implementation of class *Time*, *ExtTime* and *iWatch*.
//--

```
1   #include "iWatch.h"
2   #include <Windows.h>
3
```

```cpp
4    SYSTEMTIME GetTime()                    //get current system time
5    {
6        SYSTEMTIME sys;
7        GetLocalTime(&sys);
8        return sys;
9    }
10   //The implementation of the Time class
11   Time::Time() : hrs(0), mins(0), secs(0){}
12   Time::Time(int h, int m, int s) : hrs(h), mins(m), secs(s){}
13   Time::~Time(){}
14   void Time::Set(int h, int m, int s)
15   {
16       hrs = h; mins = m; secs = s;
17   }
18   Time& Time::operator++()
19   {
20       if (++hrs > 23)
21           hrs = 0;
22       return *this;
23   }
24   Time& Time::operator--()
25   {
26       if (--hrs < 0)
27           hrs = 23;
28       return *this;
29   }
30   void Time::Write() const
31   {
32       cout << "The time: " << hrs << ":" << mins << ":" << secs << endl;
33   }
34   int Time::Hours()
35   {
36       return hrs;
37   }
38   int Time::Minutes()
39   {
40       return mins;
41   }
42   int Time::Seconds()
43   {
44       return secs;
45   }
46   //The implementation of the *ExtTime* class
47   ExtTime::ExtTime() :Time(), zone(Zone0){}
48   ExtTime::ExtTime(int h, int m, int s, ZoneType z) : Time(h, m, s), zone(z){}
49   ExtTime::~ExtTime(){}
50   void ExtTime::Set(int h, int m, int s, ZoneType z)
```

```cpp
51  {
52      Time::Set(h, m, s);
53      zone = z;
54  }
55  ExtTime& ExtTime::operator++()
56  {
57      Time::operator++();
58      int m = zone;
59      if (++m > 23)
60          zone = Zone0;
61      zone = ZoneType(m);
62      return *this;
63  }
64  ExtTime& ExtTime::operator--()
65  {
66      Time::operator--();
67      int m = zone;
68      if (--m < 0)
69          zone = WZone1;
70      zone = ZoneType(m);
71      return *this;
72  }
73  void ExtTime::Write() const
74  {
75      string strZone[] = {"Zone0", "EZone1", "EZone2", "EZone3", "EZone4", "EZone5",
76      "EZone6", "EZone7", "EZone8", "EZone9", "EZone10", "EZone11", "EWZone12",
77      "WZone11", "WZone10", "WZone9", "WZone8", "WZone7", "WZone6", "WZone5",
78      "WZone4", "WZone3", "WZone2", "WZone1" };
79      cout << "Your zone: " << strZone[zone] << "; ";
80      Time::Write();
81  }
82  ZoneType ExtTime::Zone()
83  {
84      return zone;
85  }
86  //The implementation of the *iWatch* class
87  iWatch::iWatch() : eTime()
88  {
89      SYSTEMTIME sys = GetTime();
90      cout << "Your city: ";
91      cin >> cityName;
92      cout << "The current temperature: ";
93      cin >> temperature;
94      int zt;
95      cout << "Your zone: ";
96      cin >> zt;
97      eTime.Set(sys.wHour, sys.wMinute, sys.wSecond, ZoneType(zt));
```

```
98      }
99      iWatch::~iWatch(){}
100     void iWatch::show() const
101     {
102         cout << "Your city: " << cityName << "; The current temperature: "
103             << temperature << endl;
104     }
105     void iWatch::showTime() const
106     {
107         eTime.Write();
108     }
109     void iWatch::changeZone(int anotherZone)
110     {
111         anotherZone = anotherZone % 24;
112         int n = eTime.Zone();
113         if (n < anotherZone)
114         {
115             while (n < anotherZone)
116             {
117                 eTime.operator++();
118                 n = eTime.Zone();
119             }
120         }
121         else
122         {
123             while (n > anotherZone)
124             {
125                 eTime.operator--();
126                 n = eTime.Zone();
127             }
128         }
129     }
```

//--
//File: **example6_14.cpp**
//The program tests the *iWatch* class.
//--

```
1   #include "iWatch.h"
2
3   int main()
4   {
5       iWatch myWatch;
6       myWatch.show();
7       myWatch.showTime();
8
9       int newZone;
10      cout << "Enter a new zone ";
11      cin >> newZone;
```

```
12        myWatch.changeZone(newZone);
13        myWatch.showTime();
14        cout << "Enter a new zone ";
15        cin >> newZone;
16        myWatch.changeZone(newZone);
17        myWatch.showTime();
18
19        return 0;
20    }
```

Execute the program above, and output the following results on the screen:

```
1    Your city: Shanghai
2    The current temperature: 5
3    Your zone: 8
4    Your city: Shanghai; The current temperature: 5
5    Your zone: EZone8; The time: 21:44:4
6    Enter a new zone 18
7    Your zone: WZone6; The time: 7:44:4
8    Enter a new zone 0
9    Your zone: Zone0; The time: 13:44:4
```

Word Tips

a mere accident 只是意外	demonstrate *vt.* 演示
ambiguity *n.* 歧义，模糊	derived *adj.* 派生的
ambiguous *adj.* 不明确的	desired *adj.* 所需的
apparent *adj.* 明显的	directly *adv.* 直接地
appearance *n.* 外表	distinction *n.* 区别
application *n.* 应用	duplication *n.* 副本
calculate *vt./vi.* 计算	element *n.* 元素
calculational *adj.* 计算的	essential *adj.* 必不可少的
cast *vi./vt.* 转换	exactly *adv.* 准确地
classify *vt.* 分类	explanation *n.* 解释
collision *n.* 冲突	explicitly *adv.* 明确地
comma *n.* 逗号	extensible *adj.* 可扩展的
composition *n.* 组合	feature *n.* 特征
concept *n.* 概念	fundamental *adj.* 基本的
concrete *adj.* 具体的	further factoring 进一步分解
consultant *n.* 顾问	generate *vt.* 生成
conversion *n.* 转换	graph *n.* 图形，图表

graphically	adv.	图形化地	overriding	v. 重写
hierarchy	n.	层次，层次体系	path	n. 路径
implicitly	adv.	隐含地	plus	prep. 加
indicate	vt.	指明	preceding	adj. 前面的
indirect	adj.	间接的	rear	n. 后部
inheritance	n.	继承	recast	vt. 重变换
lattice	n.	类网格	relieve	vt. 解除
layout	n.	布局	remove	vt. 删除
leak	vt./vi.	泄漏，泄露	represent	vt. 表示
legal	adj.	合法的	secretary	n. 秘书
logical	adj.	合乎逻辑的	separate	adj. 单独的
maintainable	adj.	可维护的	specify	vt. 指定
measurable	adj.	可测量的	state	vt. 陈述
mechanism	n.	机制	straightforward	adv. 直截了当地
mere	adj.	只，仅仅	sub-object	n. 子对象
mini-payroll		小型工资	subsidy	n. 补贴
minus	n.	减号	sufficient	adj. 充足的
multiple	adj.	多的	superset	n. 超集
nonsense	adj.	毫无意义的	snippet	n. 小片段
notice	n.	注意	temporary	adj. 临时的
notion	n.	概念	unique	adj. 独一无二的
optimize	vi.	优化		

Exercises

1. Answer the following questions.

(1) If a base class has a constructor with parameters but no default values, then derived class needn't have a constructor with arguments explicitly. True or false. If *false*, please give reasons.

(2) If a base class and a derived class both include a member function with the same name, which member function will be called by an object of the derived class, assuming the scope resolution operator is not used.

(3) Assume a *Derv* class that is privately derived from class *Base*. An object of class *Derv* located in the *main* function can access

a. public members of *Derv*.

b. protected members of *Derv*.

c. private members of *Derv*.

d. public members of *Base*.

e. protected members of *Base*.

f. private members of *Base*.

(4) A class hierarchy

a. shows the same relationships as an organization chart.

b. describes "has a" relationships.

c. describes "is a kind of" relationships.

d. shows the same relationships as a family tree.

(5) Suppose that the derived class has a member object of another class. In what sequence is an object of the derived class constructed?

(6) What is the meaning of inheritance? What is the difference between composition and inheritance?

(7) Explain the difference between the ***private*** and ***protected*** members of a class.

(8) Suppose a base class and a derived class both define a function of the same name and a derived class object invokes the function. What method is called?

(9) Consider the following inheritance hierarchy:

```
class A{
private:
    int x, y;
protected:
    int z;
};
class B: public A{
private:
    int a, b, c;
};
```

(a) How many data members does *B* have? Write out these data members.

(b) Which data members in *A* are *accessed* in *B*?

2. Fill in the blanks in each of the following:

(1) A new class inherits properties from an existing class. So the relationship is called _____. A new class has an object of another class as its member. So the relationship is called _____. They are all ways of reusing existing software to create new software.

(2) When deriving a class from a base class with public inheritance, public members of the base class become _____ members of the derived class, and protected members of the base class become_____ members of the derived class. _____ part of the

class can't be accessed by the derived class.

(3) The derived class is derived from one existing class. The existing class is called _____. The members of the base class cannot be directly accessed by the derived class, but the _____ and members can be accessed by the derived class. When initializing the object of a derived class, the constructor of the _____ is invoked first.

(4) To be accessed from a member function of the derived class, the data or functions in the base class must be public or _____.

(5)
```
        class Employee {
        public:
                    _____    //constructor of class Employee
            {
                name = theName;   number = no;
            }
            void print()
            {
                cout << name << number << endl;
            }
        private:
            string name;
            int number ;
        };
        class Manager : _____    {
            //The Manager class is derived from the Employee class (public)
            int level;
        public:
            Manager(string theName, int no, int l) : _____
            {
                level=l;
            }
            void print()
            {
                _____    //call the print function of Employee
                cout << level << endl;
            }
        };
```

3. Find the error(s) in each of the following and explain how to correct it.

(1)
```
        class X{
            int a;
        public:
            X(int x) { a = x; }
```

```cpp
    void print() { cout << a; }
};
class Y: protected X{
    int b;
public:
    Y(int x, int y) { a = x; b = y; }
};
void main()
{
    Y obj(10, 2);
    obj.print();
}
```

(2)
```cpp
class baseClass{
public:
    baseClass(int a): x(a){ }
    void setX(int a){ x = a; }
private:
    int x;
};
class derivedClass: private baseClass {
public:
    derivedClass(int a, int b)
    { y = b; }
    void setXY(int a, int b)
    { x = a;   y = b; }
private:
    int y;
};
void main()
{
    derivedClass dObject(3);
    dObject.setX(50);
    dObject.setXY(50, 60);
}
```

(3)
```cpp
class Employee{
private:
    string name;
    int department;
public:
    Employee(const string& n, int d)
    { name = n; department = d; }
};
class Manager: public Employee{
```

```
        int level;
    public:
        Manager(const string& n, int d, int lvl)
        {
            name = n;
            department = d;
            level = lvl;
        }
};
```

(4)
```
class A{
    int a;
public:
    A(int x) : a(x){}
    int getA() { return a; }
};
class derived1 : private A{
public:
    derived1(int x) : A(x) {}
};
class derived2 : private A{
public:
    derived2(int x) : A(x) {}
};
class B : public derived1, public derived2{
public:
    B(int x) :derived1(x), derived2(x) {}
};
int main()
{
    B bb(20);
    bb.getA();
}
```

(5)
```
class Person{
    int ID;
public:
    void setID(int id) { ID = id; }
    int getID() const { return ID++; }
};
class Teacher: public Person {
public:
    void printTeacher() { cout << getID(); }
};
class Student: public Person{
```

```cpp
public:
    void printStudent() { cout << getID(); }
};
class TeachAssistant: public Teacher, public Student{};
void f()
{
    TeachAssistant ta;
    ta.setID(20);
    ta.printTeacher();
    ta.printStudent();
    ta.getID();
}
```

4. Write the output of the following codes.

(1)

```cpp
#include <iostream>
#include <string>
using namespace std;
class baseClass{
public:
    baseClass(string s = "", int a = 0);
    void print() const;
protected:
    int x;
private:
    string str;
};
class derivedClass: public baseClass{
public:
    derivedClass(string s = "", int a = 0, int b = 0);
    void print() const;
private:
    int y;
};
baseClass::baseClass(string s, int a)
{   str = s;   x = a;   }
void baseClass::print() const
{   cout << x << " " << str << endl;   }
derivedClass::derivedClass(string s, int a, int b) : baseClass(s, a)
{   y = b;   }
void derivedClass::print() const
{
    cout << "derived Class: " << y << endl;
    baseClass::print();
}
int main()
```

```
{
    baseClass baseO1;
    derivedClass derivedO1;
    baseO1.print();
    derivedO1.print();
    baseClass baseO2("This is the base class", 4);
    derivedClass derivedO2("ddd", 3, 9);
    baseO2.print();
    derivedO2.print();
    return 0;
}
```

(2)

```
class X{
public:
    X() { cout << "X constructing...\n"; }
    ~X() { cout << "X destroying...\n"; }
};
class Y : public X {
public:
    Y() { cout << "Y constructing...\n"; }
    ~Y() { cout << "Y destroying...\n"; }
};
class Z : public Y{
public:
    Z() { cout << "Z constructing...\n"; }
    ~Z() { cout << "Z destroying...\n"; }
};
int main(int argc, char* argv[])
{
    Z zz;
    return 0;
}
```

(3)

```
class A{
    int pA;
public:
    A(int a) { pA = a; cout << "constructing class A" << endl; }
    int getpA() { return pA; }
};
class B: virtual public A {
public:
    B(int a): A(a){ cout << "constructing class B" << endl;}
    void OnB() { cout << "B: " << getpA() << endl;}
};
```

```cpp
class C: public B, virtual public A {
public:
    C(int a):A(a), B(a) { cout << "constructing class C" << endl; }
    void OnC(){ cout << "C: " << getpA() + 1 << endl; }
};
int main()
{
    C cc(5);
    cc.OnB();
    cc.OnC();
    return 0;
}
```

(4)
```cpp
class base{
    int a;
public:
    base(int sa)
    { a=sa;
      cout << "base" << endl;  }
};
class base1:virtual public base {
    int b;
public:
    base1(int sa,int sb):base(sa)
    { b=sb;
      cout << "base1" << endl;  }
};
class base2: virtual public base{
    int c;
public:
    base2(int sa, int sc):base(sa)
    { c=sc;
      cout << "base2" << endl;  }
};
class derived:public base1, public base2{
    int d;
public:
    derived(int sa, int sb, int sc, int sd) : base(sa), base1(sa,sb), base2(sa,sc)
    { d=sd;
      cout << "derived" << endl;  }
};
void main()
{
    derived obj(2, 4, 6, 8);
}
```

5. The UML diagram of a class structure is shown in Figure 6-14.

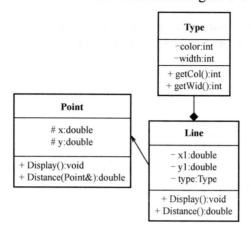

Figure 6-14　The UML diagram of the classes

(1) Explain the relationship among classes *Point*, *Line* and *Type*.

(2) Explain the meanings of operators #, -, and + in the class diagram.

(3) Define class *Line*. (Hint: The class only contains the declarations of all member functions.)

(4) How many data members does class *Line* have? Write out them.

(5) Define the constructor of class *Line*.

(6) Define function *Display* in class *Line*. (Hint: the function displays all information of data members in class *Line*.)

(7) Write the *main* function to test class *Line*.

6. Design and implement a mini-payroll system including three different people: *Teacher*, *Student* and *TeachAssistant*. This class hierarchy is described graphically in Figure 6-15.

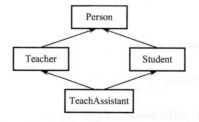

Figure 6-15　The class hierarchy diagram

(1) Class *Person* has data members: *number*, *name* and *monthly salary* and member functions to calculate the monthly salary and to display all information.

(2) Classes *Teacher* and *TeachAssistant* both have another data member—*level*.

(3) The calculation of the monthly salary is defined as follows:

The *Teacher*'s monthly salary is determined by 50 Yuan per working hour and 1500 Yuan subsidy.

The *Student* gets 300 Yuan subsidy per month;

The *TeachAssistant* is a part-time student and has a teacher's benefits. The salary is the student's subsidy, plus an extra salary of 10 Yuan per working hour.

7. Derive a class called *Employee2* from the *Employee* class in this chapter. This new class should add a *double* type data item called *compensation*, and also an *enum* type called *period* to indicate whether the employee is paid hourly, weekly, or monthly. You will derive the *Manager*, *Scientist*, and *Laborer* classes from *Employee2* instead of *Employee*. However, note that in many circumstances it might be more in the spirit of OOP to create a separate base class called *Compensation* and three new classes *Manager2*, *Scientist2*, and *Laborer2*, and use multiple inheritance to derive these three classes from the original *Manager*, *Scientist*, and *Laborer* classes and from *Compensation*. In this way, none of the original classes needs to be modified.

8. Write a program.

(1) Define a *Vehicle* class that has a constructor with parameters and protected data members *wheels* (number of wheels) and *weight* (weight of vehicle).

(2) Class *Car* is derived from *Vehicle* privately and contains the data member *passenger_load* (largest number of passengers).

(3) Class *Truck* is derived from *Vehicle* privately and contains the data members *passenger_load* (largest number of passengers) and *pay_load* (load capacity)

(4) Test the three classes by using the *main* function.

9. Write a program to implement the management of book and magazine sales. The requirements are as follows:

(1) Input the sales records of books and magazines.

(2) Output the names of books and magazines and their sales situations. If the situation is good, this means that the book sales are more than 500 Yuan monthly, and the magazine sales more than 2500 Yuan; otherwise the situation is poor.

Chapter 7   Polymorphism and Virtual Functions

> *General propositions do not*
> *decide concrete cases.*
> *—Oliver Wendell*

Objectives
- What polymorphism is, how it makes programming more convenient, and how it makes systems more extensible and maintainable
- To declare and use virtual functions to effect polymorphism
- The distinction between abstract and concrete classes
- To declare pure virtual functions to create abstract classes

7.1   Polymorphism

7.1.1   Introduction to Polymorphism

The key idea behind OOP is polymorphism. Polymorphism is derived from a Greek word meaning "many forms". We speak the types related by inheritance as polymorphic types, because in many cases we can use the "many forms" of a derived or base type interchangeably.

In programming languages, *polymorphism* means that some codes or operations or objects behave differently in different contexts. For example, the + operator in C++ can implement the addition operation with different data or different types.

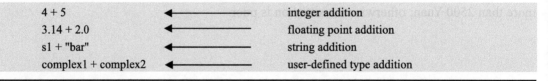

Polymorphism is a key feature of object-oriented programming that allows the values of different data types to be handled by using a uniform interface.
多态性是一个面向对象程序设计的关键特性，它允许使用统一的接口处理不同数据类型。

The purpose of *polymorphism* is to implement a style of programming called *message-passing* in the literature, in which the objects of various types define a common interface of operations for users.

Typically, polymorphism occurs when there is a class hierarchy in which the classes are related by inheritance.

7.1.2 Binding

When a C++ program is executed, it executes sequentially, beginning at the top of function *main*. When a function call is encountered, the point of execution jumps to the initial point of the function being called. How does the CPU know to do this?

When a program is compiled, the compiler converts each statement in your C++ program into one or more lines of machine language. Each line of machine language is given its own unique sequential address. This is no different for functions—when a function is encountered, it is converted into machine language and given the next available address. Thus, each function ends up with a unique machine language address.

> **Binding** is a process that is used to convert identifiers (such as variable and function names) into machine language addresses. In other words, **binding** is an association, such as between identifiers (variables or function names) and operations (a type or the specific function body).
>
> There are two types of binding according to the time at which a binding takes place, i.e. **static binding** and **dynamic binding**.
>
> 绑定是一种用于将标识符（如变量和函数名）转换成机器语言地址的过程。换句话说，**绑定**是一种关联，例如在标识符（变量或函数名）和操作（类型或特定函数体）之间的关联。
>
> 根据绑定发生的时间，有两种类型的绑定，即**静态绑定**和**动态绑定**。

Static Binding

By default, C++ matches a function call with the correct function definition at compile time. This is called ***static binding***. This kind of binding is also known as ***compile-time binding***. For example, the operator overloading and the function overloading mentioned in the previous chapters are that the compiler associates the operator functions and overloaded functions with their definitions at compile-time.

Each function has a unique address. Thus, when the compiler encounters a function call, it replaces the function call with a machine language instruction that tells the CPU to jump to the address of the function.

Example 7-1: Static binding of member functions *f* within the classes.

```
//-------------------------------------------------------------------------
//File: example7_1.cpp
//This program illustrates how to implement static binding of member functions f within the classes.
//-------------------------------------------------------------------------
1   #include <iostream>
2   using namespace std;
3
4   class baseClass {
```

```
 5   public:
 6       void f() { cout << "Class baseClass" << endl; }
 7   };
 8
 9   class derivedClass: public baseClass {
10   public:
11       void f() { cout << "Class derivedClass" << endl; }
12   };
13
14   void fn(baseClass& arg) {
15       arg.f();
16   }
17
18   int main()
19   {
20       baseClass one;
21       fn(one);
22       derivedClass two;
23       fn(two);
24       return 0;
25   }
```

Result:

```
1  Class baseClass
2  Class baseClass
```

In Example 7-1, the *fn* function in Line 14 has a formal reference parameter of class *baseClass*. We can call the *fn* function by using an object of either class *baseClass* or class *derivedClass* as an argument. Let's look at the results generated by the statements in Lines 21 and 23. The results show only the output of *baseClass*, even though in these statements a different class object is passed as a parameter. When object *two* is passed as a parameter to function *fn*, we would expect that the output should be "Class derivedClass". However, actual output is "Class baseClass". Why is this? This is because the binding of member function *f*, in the body of function *fn*, occurred at compile time. Because the formal parameter *arg* of function *fn* is of class *baseClass*, for statement *arg.f()* in Line 15, the compile associates function *f* with class *baseClass* at compile time.

Dynamic Binding

When the compiler matches a function call with the correct function definition at run-time, this is called ***dynamic binding***. The code associated with the procedure does not know until the program is executed. This kind of binding is also known as ***run-time binding***. You can declare a function with the ***virtual*** keyword if you want the compiler to use dynamic binding for that specific function.

The statement in Line 23 of Example 7-1 indicates that the actual parameter is of class *derivedClass*. Thus, when the body of function *fn* executes, logically the *f* function of object *two* should be executed, which is not the case. How does C++ correct this problem? To generate the output we would expect, the example above is somewhat changed in Line 6.

Example 7-2: Dynamic binding of member functions *f* within the classes.

```
//----------------------------------------------------------------------------
//File: example7_2.cpp
//This program illustrates how to implement dynamic binding of member functions f within
//the classes.
//----------------------------------------------------------------------------
1  #include <iostream>
2  using namespace std;
3
4  class baseClass {
5  public:
6      virtual void f() { cout << "Class baseClass" << endl; }
7  };
8
9  class derivedClass: public baseClass {
10 public:
11     void f() { cout << "Class derivedClass" << endl; }
12 };
13
14 void fn(baseClass& arg) {
15     arg.f();
16 }
17
18 int main()
19 {
20     baseClass one;
21     fn(one);
22     derivedClass two;
23     fn(two);
24     return 0;
25 }
```

Result:

```
1  Class baseClass
2  Class derivedClass
```

The result is desirable. This is because the mechanism of ***virtual function*** is provided in

Example 7-2. The binding of virtual function occurs at program execution time, but not at compile time.

The two kinds of binding both have the idea of polymorphism. They are referred to as *compile-time polymorphism* and *run-time polymorphism*.

Think These Over

1. What is polymorphism?
2. What is the difference between static binding and dynamic binding?

7.2 Virtual Functions

A virtual function is a member function of the base class and is redefined by the derived class. The compiler will guarantee the correct correspondence between objects and the functions applied to them. A virtual function is created using the keyword *virtual* which precedes the name of the function.

7.2.1 Definition of Virtual Functions

For example, suppose a graphics program includes several different shapes: a cube and a cuboid, and so on, as shown in Figure 7-1.

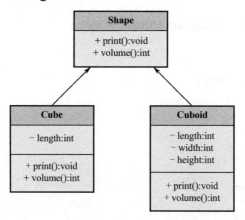

Figure 7-1   The hierarchy of classes *Shape*, *Cube* and *Cuboid*

Each of these classes has a member function *volume* that causes the object's volume to be calculated. We define a *Shape* class according to the analysis of a class hierarchy in Figure 7-1.

```
class Shape{
public:
    Shape();
    virtual int volume();
};
```

To enable this kind of behavior, we declare function *volume* in the base class as a ***virtual function***, and we override function *volume* in each of the derived classes to calculate the volume of appropriate shape.

When a derived class inherits the class containing the virtual function, it has the ability to redefine the virtual functions. A virtual function has different functionality in the derived class. The virtual function implements the philosophy of one interface and multiple methods.

> A **virtual function** is a member function of the base class and is redefined by the derived class. The virtual function within the base class provides the form of the interface to the function.
> 虚函数是基类的成员函数，并由派生类重新定义。基类中的虚函数提供了函数的接口形式。

Virtual functions can be accessed by using a base class pointer. A pointer to the base class can be created. A base class pointer can contain the address of the derived object as the derived object contains the subset of a base class object. Every derived class is also a base class. When a base class pointer contains the address of the derived class object, at run time, which version of the virtual function being called depends on the type of object contained by the pointer. Here is a program which illustrates the working of virtual functions.

Example 7-3: Definition of the class with virtual member functions.

```
//-----------------------------------------------------------------------
//File: example7_3.cpp
//This program illustrates the definition of a virtual function within class Shape.
//-----------------------------------------------------------------------
1   #include<iostream>
2   using namespace std;
3   class Shape{
4   public:
5       Shape(){ }
6       virtual int volume()
7       {
8           cout << "Virtual function of the base class " << endl;
9           return 0;
10      }
11  };
12
13  class Cube: public Shape{
14  public:
15      Cube(int l):Shape(), length(l) {}
16      virtual int volume()
17      {
18          cout <<"The volume of the cube is" ;
19          return length * length * length;
20      }
21  private:
22      int length;
```

· 217 ·

```
23     };
24     class Cuboid: public Shape {
25     public:
26         Cuboid(int l, int w, int h):Shape(), length(l), width(w), height(h){}
27         virtual int volume()
28         {
29             cout << "The volume of the cuboid is ";
30             return length * width * height;
31         }
32     private:
33         int length;
34         int width;
35         int height;
36     };
37     void fn(Shape* s)              //a top-level function
38     {
39         cout << s->volume() << endl;
40     }
41     int main()
42     {
43         Cube c1(10);
44         Cuboid c2(15,16,20);
45         fn(&c1);
46         fn(&c2);
47         return 0;
48     }
```

Result:

```
1   The volume of the cube is 1000
2   The volume of the cuboid is 4800
```

The program has base class *Shape* and derived classes *Cube* and *Cuboid* which inherit base class *Shape*. The base class defines a virtual function *volume*.

Line 6 defines the virtual function *volume* of base class *Shape*. The **virtual** keyword precedes the function returning-value type. In Line 16, derived class *Cube* is derived from base class *Shape* redefines the virtual function *volume*. Likewise, in Line 27, derived classes *Cuboid* also redefines function *volume*. In Line 37, the statement

    void fn(Shape* s);

declares the *fn* function with the parameter of a *s* pointer to base class *Shape*. In general, the *fn* is called a ***top-level*** function. In Lines 43 and 44, the statements declare objects c1 and c2 as classes *Cube* and *Cuboid*. In Line 45, the statement

    fn (&c1);

states that the address of actual parameter c1 is passed to pointer *s* when the program executes the statement in Line 39, the statement

```
cout << s -> volume() << endl;
```

calls the *volume* virtual function of derived class *Cube*. The *s* pointer contains the address of object *c*1 of derived class *Cube*. Likewise, the statement in Line 46 states that the address of actual parameters *c*2 is passed to pointer *s* of base class *Shape*. The *s* pointer contains the address of the object of derived class *Cuboid*.

When a virtual function is not defined by the derived class, the version of the virtual function defined by the base class is called. When a derived class contains the virtual function inherited from another class, the virtual function can be overloaded by the new derived class. This means that the virtual function can be inherited.

Once a function is declared as virtual, it remains virtual all the way down the inheritance hierarchy from that point, even if that the function is not explicitly declared as virtual when a class overrides it. If the member function in the derived class is virtual, the keyword virtual can be omitted in the derived class. But the keyword virtual cannot be omitted in the base class.

一旦一个函数被声明为 virtual，它在派生类中一直保持 virtual，即使该函数在类重写时未显式声明为 virtual。如果派生类中的成员函数是虚的，在派生类中，关键字 virtual 可以被省略。但是，在基类中，关键字 virtual 不能被省略。

7.2.2 Extensibility

If the *volume* function is defined as **virtual** in the base class, you can add as many new class types as you want without changing the *fn* function. In a well-designed OOP program, most or all your functions will follow the model of function *fn* and communicate only with the base-class **interface**. Such a program is **extensible** because you can add new functionality by inheriting new data types from the common base class. The functions that manipulate the base class interface will not need to be changed at all to accommodate the new classes. Here is the *Shape* example with more virtual functions and a number of new classes, all of which work correctly with the old, unchanged *fn* function. The extensibility of class hierarchy mentioned in Example 7-3 is shown in Figure 7-2.

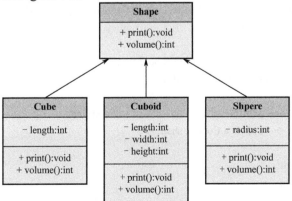

Figure 7-2　Extensibility of class *Shape*

Example 7-4: Extensibility of class *Shape*.

```
//--------------------------------------------------------------------
//File: example7_4.cpp
//This program extends a new Sphere class without changing the fn function.
//--------------------------------------------------------------------
1   #include <iostream>
2   using namespace std;
3   class Shape{
4   public:
5       Shape(){ }
6       virtual int volume()                        //interface
7       {
8           cout << "The volume of the shape is none" << endl;
9           return 0;
10      }
11      virtual viod print()                        //interface
12      { cout << "This is a common shape\n"; }
13  };
14
15  class Cube: public Shape{
16  public:
17      Cube(int l): Shape(), length(l){}
18      virtual int volume()
19      {
20          cout << "The volume of the cube is " ;
21          return length * length * length;
22      }
23      virtual viod print()                        //interface
24      { cout << "his is a Cube shape\n"; }
25  private:
26      int length;
27  };
28  class Cuboid: public Shape{
29  public:
30      Cuboid(int l, int w, int h): Shape(), length(l), width(w), height(h){}
31      virtual int volume()
32      {
33          cout << "The volume of the cuboid is ";
34          return length * width * height;
35      }
36      virtual viod print()                        //interface
37      { cout << "This is a Cuboid shape\n"; }
38  private:
39      int length;
40      int width;
41      int height;
```

```
42    };
43    class Sphere: public Shape{
44    public:
45        Sphere(int r): Shape(), radius(r){}
46        virtual int volume()
47        {
48            cout <<" The volume of the sphere is " ;
49            return (int)3 * 3.14 * radius * radius * radius / 4;
50        }
51        virtual viod print()                     //interface
52        {   cout << "This is a Sphere shape\n";   }
53    private:
54        int radius;
55    };
56    void fn(Shape* s)
57    {
58        s->print();
59        cout << s->volume() << endl;
60    }
61    int main()
62    {
63        Cube c1(10);
64        Cuboid c2(15,16,20);
65        Sphere sp(30);
66        fn(&c1);
67        fn(&c2);
68        fn(&sp);
69        return 0;
70    }
```

Result:

```
1    This is a Cube shape
2    The volume of the cube is 1000
3    This is a Cuboid shape
4    The volume of the cuboid is 4800
5    This is a Sphere shape
6    The volume of the sphere is 63585
```

7.2.3 Principle of Virtual Functions

How can *dynamic binding* happen by virtual functions? All the work goes on behind the scenes by the compiler, which installs the necessary dynamic binding mechanism when you ask it to (you are asked to create virtual functions). Because programmers often benefit from understanding the mechanism of virtual functions in C++, this section will elaborate on the way the compiler implements this mechanism. The *virtual* keyword tells the compiler that it should not perform static binding. Instead, it should automatically install all the mechanisms necessary to perform dynamic binding. This means that if you call functions *volume* of derived class objects through an address of the *Shape* base class, you will get the proper function.

To accomplish this, the typical compiler creates a single **virtual function table** (called the **VTABLE**) for each class that contains **virtual** functions. The compiler places the addresses of the virtual functions for that particular class in the VTABLE. In each class with virtual functions, it secretly places a pointer, called the **vpointer** (abbreviated as VPTR), which points to the VTABLE for that object, as shown in Figure 7-3. When you make a virtual function call through a base class pointer (that is, when you make a polymorphic call), the compiler quietly inserts code to fetch the VPTR and look up the function address in the VTABLE, thus calling the correct function and causing late binding to take place.

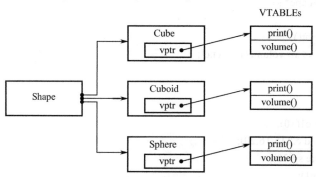

Figure 7-3   Virtual tables of Example 7-4

In Example 7-4, all of these—setting up the VTABLE for each class of *Cube*, *Cuboid* and *Shape*, initializing the VPTR, inserting the code for the virtual function call—happen automatically, so you do not have to worry about it. With virtual functions, the proper function is called for an object, even if the compiler cannot know the specific type of the object. Now we test the sizes of the *Cube* and *Cuboid* objects as follows:

```
int main()
{
    Cube c1(10);
    cout << "size of c1" << sizeof(c1) << endl;
    Cuboid c2(10, 20, 30);
    cout << "size of c2" << sizeof(c2) << endl;
    return 0;
}
```

The results are

    size of c1 8
    size of c2 16

Actually, the sizes of the *Cube* and *Cuboid* objects are 4 and 12. But the *Cube* and *Cuboid* classes contain a *vptr* virtual point due to the virtual function. The *vptr* size is 4. Therefore, the results above are obtained.

7.2.4 Virtual Destructors

One thing recommended for the class with pointer member variables is that the class should have the destructor (see Section 4.6.2). The destructor is automatically executed when the class object goes out of the scope. Then, if the object creates dynamic objects, the destructor can be designed to deallocate the storage for them.

The base class destructor should always be virtual. Suppose you use the ***delete*** operator with a base class pointer to a derived class object to destroy the derived class object. If the base class destructor is not virtual, but like a normal member function, then the ***delete*** operator calls the destructor for the base class, but not the destructor for the derived class. This will cause only the base part of the object to be destroyed. The example 7-5 program shows how this looks.

Example 7-5: Tests non-virtual destructors.

```
//------------------------------------------------------------
//File: example7_5.cpp
//This program defines a Base class with a non-virtual destructor.
//------------------------------------------------------------
1    #include <iostream>
2    using namespace std;
3
4    class Base{
5    public:
6        ~Base()              //non-virtual destructor
7        { cout << "Base destroyed\n"; }
8    };
9
10   class Derived : public Base{
11   public:
12       ~Derived()
13       { cout << "Derived destroyed\n"; }
14   };
15
16   int main()
17   {
18       Base* pBase = new Derived;
19       delete pBase;
20       return 0;
21   }
```

Result:

```
1 Base destroyed
```

This shows that the destructor for the *Derived* part of the object is not called. In the program, the base class destructor is not virtual, but you can make it so by changing the definition of the Base class destructor,

```
class Base{
public:
    virtual ~Base()                //virtual destructor
    { cout << "Base destroyed\n"; }
};
```

Now the result is

```
1 Derived destroyed
2 Base destroyed
```

Now both parts of the *derived* class object are destroyed properly.

If none of the destructors has anything important to do (like deallocating memory obtained with **delete**), then virtual destructors aren't necessary. But, in general, to ensure that derived class objects are destroyed properly, you should make virtual the destructors in all base classes.

如果不需要使用析构函数做任何重要的事情（如使用 **delete** 释放内存），则没必要将析构函数声明为虚函数。但是，一般来说，为了确保派生类对象能被确切地析构，应该让所有基类的析构函数为虚函数。

7.2.5 Function Overloading and Function Overriding

We have introduced function overloading and function overriding in the previous sections. These terms are similar, and they do similar things.

When you ***override*** a member function of the class, you create a member function in a derived class with the same name as a function in the base class and the ***same signature***. If we define a function as virtual in the base class, the function with the same name and same signature but no keyword virtual in the derived class is also the virtual function because of ***overriding***.

When you ***overload*** a function, you create more than one function with the same name, but with a ***different signature***. If we define a function as virtual in the base class, the function with the same name and different signature in the derived class is **not** a virtual function. We call this ***overloading***.

Example 7-6: Function overloading and function overriding in inheritance.

```
//-------------------------------------------------------------------------
```

//File: **example7_6.cpp**
//This program illustrates the difference between functions overloading and overriding.
//--

```cpp
1   #include <iostream>
2   #include <string>
3   using namespace std;
4   class Base {
5   public:
6     virtual int f() const
7     {
8         cout << "Base::f()\n";
9         return 1;
10    }
11    virtual void f(string) const { cout << "function f has a string parameter\n"; }
12    virtual void g() const { cout << " Base::g()\n"; }
13  };
14  class Derived1 : public Base {
15  public:
16    void g() const
17    { cout << "Derived1::g()\n"; }
18  };
19  class Derived2 : public Base {
20  public:
21    //Overriding a virtual function
22    int f() const
23    { cout << "Derived2::f()\n";   return 2; }
24  };
25  class Derived3 : public Base {
26  public:
27    //Changing return type
28    //void f() const              //error: overriding virtual function differs from Base::f()
29    //{ cout << "Derived3::f()\n"; }
30  };
31  class Derived4 : public Base {
32  public:
33    //Overloading the f function by changing parameter list
34    int f(int) const
35    { cout << "Derived4::Overloading f()\n";   return 4; }
36    int f() const
37    { cout << "Derived4:: Overriding f()\n";   return 4; }
38
39  };
40  int main()
41  {
42    string s("hello");
43    int x;
44
```

```
45      cout << "Create the Derived1 object.\n";
46      Derived1 d1;
47      x = d1.f();
48      d1.f(s);
49      d1.g();
50
51      cout << "Create the Derived2 object.\n";
52      Derived2 d2;
53      x = d2.f();
54
55      cout << "Create the Derived4 object.\n";
56      Derived4 d4;
57      x = d4.f(1);
58
59      Base& br = d4; //Upcast
60      x = br.f();
61      br.f(s);        //Base version available
62      return 0;
63   }
```

Result:

```
1    Create the Derived1 object.
2    Base::f()
3    function f has a string parameter
4    Derived1::g()
5    Create the Derived2 object.
6    Derived2::f()
7    Create the Derived4 object.
8    Derived4:: Overloading f()
9    Derived4::Overriding f()
10   function f has a string parameter
```

The result may not be what you expect. Let us look first at class *Derived*1. There are two virtual functions *f* and a virtual function *g* defined within class *Base*. Since class *Derived*1 directly inherits two functions *f* from class *Base*, the output of the statements in Lines 47 and 48 are:

```
Base::f()
function f has a string parameter
```

However, the *g* member function of class *Dervied*1 overrides the one of its base class. When the statement in Line 49 executes, the *g* member function of *Dervied*1 is called. The output is

```
Derived1::g()
```

Similarly, because virtual function *f*, without the parameter of class *Derived*2, overrides

• 226 •

the one of its base class, the output of the statement in Line 53 is

> Derived2::f()

Class *Derived*4 defines another member function *f*, with different signatures from the virtual functions *f* of its base class. Consequently, when the statement in Line 57 executes, object *d*4 calls overloading function *f* of *Derived*4 by a parameter match, rather than the *f* function of class *Base*. The output is

> Derived4::f()

From this point, overriding is useful when you inherit from a base class and wish to extend or modify its functionality. Even when object *d*4 is cast as the base class in Line 59, it calls overridden function *f* of *Derived*4 in Line 60, but not the base one.

Think These Over

1. What are the advantages of using virtual functions in inheritance?
2. What is the difference between function overloading and function overriding?

7.3 Abstract Base Classes

Often in a design, you want the base class to present only an interface for its derived classes. That is, you don't allow anyone to create an object of the base class, but only want to upcast to it so that its interface can be used. This is accomplished by making that class ***abstract***.

Abstract base classes act as the expressions of general concepts from which more specific classes can be derived.

If an ***abstract base class*** cannot be instantiated, it exists extensively for inheritance and it must be inherited. There are scenarios in which it is useful to define a class that is not intended to be instantiated because such classes normally are used as base-classes in inheritance hierarchies.

This class must be inherited. This class is mostly used as a base class. You cannot create an object of an abstract class type; however, you can use pointers and references to abstract class types.

A class that contains at least one ***pure virtual function*** is considered as an abstract class. The classes derived from the abstract class must implement the pure virtual function or they, too, are abstract classes.

> A **pure virtual function** is a function which contains no definition in the base class. You declare a pure virtual function by using a pure specifier (= 0) in the declaration of a virtual member function in the class declaration.
>
> 纯虚函数是一个在基类中没有定义的函数。在类声明时使用 pure 说明符（= 0）将一个虚成员函数声明为纯虚函数。

The general form of a virtual function is

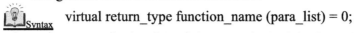   virtual return_type function_name (para_list) = 0;

where *return_type* is the type of the returning value, *function_name* is the name of the virtual function and *para_list* is the parameter list.

Each derived class should contain the definitions of virtual functions. If the derived class does not define the virtual function, the compiler will return an error.

> A class which contains one or more pure virtual function is called an **abstract base class**. Abstract base classes act as expressions of general concepts from which more specific classes can be derived.
> 包含一个或多个纯虚函数的类称为**抽象基类**。抽象基类作为通用概念的表示，从中可以导出更多具体的类。

The following program illustrates how to define a pure virtual function.

Example 7-7: Definition of an abstract base class.

```
//-----------------------------------------------------------------------
//File: example7_7.cpp
//This program defines a Shape class with a pure virtual function.
//-----------------------------------------------------------------------
1  #include<iostream>
2  using namespace std;
3  class Shape{
4  public:
5      Shape(){ }
6      virtual int volume() = 0;              //pure virtual function
7  };
8  class Cube: public Shape{
9  public:
10     Cube(int l): Shape(), length(l){}
11     virtual int volume()
12     {
13         cout << "Volume function of Cube " << endl;
14         return length * length * length;
15     }
16 private:
17     int length;
18 };
19 class Cuboid: public Shape{
20 public:
21     Cuboid(int l, int w, int h) : Shape(), length(l), width(w), height(h){}
22     virtual int volume()
23     {
24         cout << "Volume function of Cuboid " << endl;
25         return length * width * height;
26     }
27 private:
28     int length;
```

```
29      int width;
30      int height;
31   };
32   int main()
33   {
34      Shape *s;
35      //Shape s1;              //error: abstract class cannot declare objects
36      Cube c1(10);
37      Cuboid c2(15,16,20);
38      s = &c1;
39      cout << "The volume of the cube " << s->volume() << endl;
40      s = &c2;
41      cout << "The volume of the cuboid " << s->volume() << endl;
42      return 0;
43   }
```

Result:

```
1   Volume function of Cube
2   The volume of the cube 1000
3   Volume function of Cuboid
4   The volume of the cuboid 4800
```

In Line 6 of Example 7-7, the statement

```
virtual int volume() = 0;
```

declares the pure virtual function *volume* which has no definition. In Line 35, the statement:

```
Shape s1;
```

is wrong because the objects of the abstract class *Shape* cannot be created.

The statement in Line 6 tells the compiler to reserve a slot for a function in the VTABLE, but not to put an address in that particular slot. Even if only one function in a class is declared as pure virtual, the VTABLE is incomplete.

If the VTABLE for a class is incomplete, what is the compiler supposed to do when someone tries to make an object of that class? It cannot safely create an object of an abstract class, so you get an error message from the compiler. Thus, the compiler guarantees the purity of the abstract class. By making a class abstract, you can ensure that the client programmer cannot misuse it.

Pure virtual functions are helpful because they make explicit the abstractness of a class and tell both the user and the compiler how it was intended to be used.

Note that pure virtual functions prevent an abstract class from being passed into a function by value. Thus, it is also a way to prevent ***object slicing*** (which will be described shortly). By making a class abstract, you can ensure that a pointer or reference is always used during upcasting to that class.

The objects of an abstract base class cannot be created as it does not contain the definition of one or more member functions except virtual functions. The pointers to an abstract class can be created and the references to the abstract class can be made.

不能创建抽象基类的对象，因为它不包含除虚函数外的一个或多个成员函数的定义。可以创建抽象类的指针，也可以声明抽象类的引用。

 Think These Over

1. What is the difference between abstract base classes and virtual base classes?
2. Why is the abstract class objects not declared?
3. How to understand inheritance interface and inheritance implementation?

7.4 Case Study: A Mini-System

We design a mini-system that includes different types of people in the university. There are five classes, that is, *Person, Student, Employee, Faculty* and *Staff*, to be defined. A class named *Person* has two derived classes named *Student* and *Employee*. Make the derived classes *Faculty* and *Staff* of *Employee*. A person has a name, address, phone number and e-mail address. A student has a class status (freshmen, sophomore, junior, or senior). An employee has an office, salary, and date-hired. Define a class named *Date* that contains the fields year, month and day. A faculty member has office hours and a rank. A staff member has a title. Define a constant virtual *toString* function in the *Person* class and override it in each class to display the class name and the person's name. The UML diagram of these six classes is described in Figure 7-4.

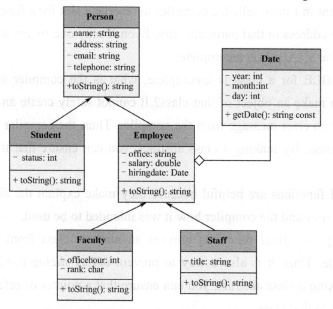

Figure 7-4 The UML diagram of six classes

· 230 ·

Example 7-8: A mini-system with six classes.

```
//----------------------------------------------------------------------------
//File: people.h
//This program defines six types of people.
//----------------------------------------------------------------------------
1   class Date{
2   public:
3       Date(int y, int m, int d);
4       Date(const Date& d);
5       string getDate() const;
6   private:
7       int year, month, day;
8   };
9
10  class Person{
11  public:
12      Person(string strN, string strA, string strP, string strEm);
13      virtual string toString();
14  private:
15      string name, address, phone, email;
16  };
17
18  class Student: public Person{
19  public:
20      Student(int st, string strN, string strA, string strP, string strEm);
21      string toString();
22  private:
23      int status;
24  };
25
26  class Employee: public Person{
27  public:
28      Employee(string strOf, double sal, Date date, string strN, string strA,
29              string strP, string strEm);
30      string toString();
31  private:
32      string office;
33      double salary;
34      Date hiringDate;
35  };
36
37  class Faculty : public Employee{
38  public:
39      Faculty(int ohour, char r, string strOf, double sal, Date date, string strN,
40              string strA, string strP, string strEm);
41      string toString();
```

```
42      private:
43          int officeHour;
44          char rank;
45      };
46
47      class Staff: public Employee{
48      public:
49          Staff(string t, string strOff, double sal, Date date, string strN, string strA,
50              string strP, string strEm);
51          string toString();
52      private:
53          string title;
54      };
```

//---
//File: **people.cpp**
//The program defines the member functions of all classes.
//---

```
1   #include <iostream>
2   #include <string>
3   using namespace std;
4   #include <string.h>
5   #include "people.h"
6
7   Date::Date(int y, int m, int d)
8   { year = y; month = m; day = d; }
9   Date::Date(const Date& d)
10  {
11      year = d.year;
12      month = d.month;
13      day = d.day;
14  }
15  string Date::getDate() const
16  {
17      string strDate;
18      char temp[10];
19      sprintf(temp, "%d-%d-%d", year, month, day);
20      strDate = temp;
21      return strDate;
22  }
23  //definition of the person class
24  Person::Person(string strN, string strA, string strP, string strEm) : name(strN),
25          address(strA), phone(strP),email(strEm){}
26  string Person::toString()
27  {
28      return "Name: "+name+"; Address: "+address+"; Phone: "+phone+"; Email: "+email;
29  }
30  //definition of the student class
```

```cpp
31   Student::Student(int st, string strN, string strA, string strP, string strEm):
32            Person(strN, strA, strP, strEm), status(st){}
33   string Student::toString()
34   {
35       string tempStatus;
36       switch(status){
37       case 1:
38            tempStatus = "Freshman";
39            break;
40       case 2:
41            tempStatus = "sophomore";
42            break;
43       case 3:
45            tempStatus = "Junior";
46            break;
47       default:
48            tempStatus = "Senior";
49       }
50       cout <<"Student Information\n";
51       return Person::toString() + " I am " + tempStatus;
52   }
53   //definition of the employee class
54   Employee::Employee(string strOf, double sal, Date date, string strN, string strA, string strP,
55       string strEm) : Person(strN, strA, strP, strEm), office(strOf), salary(sal), hiringDate(date){}
56   string Employee::toString()
57   {
58       char ch[20];
59       sprintf(ch, "Salary: %6.2f", salary);
60       string temp = ch;
61       return Person::toString() + temp + "; Hiring date: " + hiringDate.getDate();
62   }
63   //definition of the faculty class
64   Faculty::Faculty(int ohour, char r, string strOf, double sal, Date date, string strN, string strA,
65                   string strP, string strEm): Employee(strOf, sal, date, strN, strA, strP, strEm),
66                   officeHour(ohour), rank(r){}
67   string Faculty::toString()
68   {
69       char ch[20];
70       sprintf(ch, "; Office hour: %d", officeHour);
71       string temp = ch;
72       cout << "Faculty Information\n";
73       return Employee::toString() + ch + "; Rank: " + rank;
74   }
75   //definition of the staff class
76   Staff::Staff(string t, string strOff, double sal, Date date, string strN, string strA, string strP,
77                   string strEm): Employee(strOff, sal, date, strN, strA, strP, strEm), title(t){}
78   string Staff::toString()
```

```
79    {
80        cout << "Staff Information\n";
81        return Employee::toString() + "; Title: " + title;
82    }
```

//--
//File: **example7_8.cpp**
//This program tests these classes through dynamic polymorphism.
//--

```
1   #include <iostream>
2   #include <string>
3   using namespace std;
4   #include "people.h"
5
6   void test(Person& p)
7   {
8       cout << p.toString() <<endl;
9   }
10  int main()
11  {
12      Date date(2015, 05, 23);
13
14      Student st1(2, "Li Ming", "Shenliao Road, Shenyang", "1800403456", "11234@qq.com");
15      test(st1);
16
17      Faculty fa(20, 'A',"210", 9000.80, date, "Mary", "Shenliao Road, Shenyang",
18              "1860774603", "12378@qq.com");
19      test(fa);
20
21      Staff st("administrator", "301", 5800.00, date, "Zhang", "Shenliao Road, Shenyang",
22              "1390000000", "11234678@qq.com");
23      test(st);
24      return 0;
25  }
```

Result:

```
1   Student Information
2   Name: Li Ming; Address: Shenliao Road, Shenyang; Phone: 1800403456;
    Email: 11234@qq.com I am sophomore
3   Faculty Information
4   Name: Mary; Address: Shenliao Road, Shenyang; Phone: 1860774603;
    Email: 12378@qq.com Salary: 9000.80; Hiring date: 2015-5-23; Office hour: 20; Rank: A
5   Staff Information
6   Name: Zhang; Address: Shenliao Road, Shenyang; Phone: 1390000000;
    Email: 11234678@qq.com Salary: 5800.00; Hiring date: 2015-5-23; Title: administrator
```

Word Tips

argument   *n.*   参数
binding   *n.*   绑定
compile   *vt.*   编译
context   *n.*   上下文
convert   *vi.*   转变，转化
denote   *vt.*   表示
display   *vt.*   显示
dynamic binding   动态绑定
execute   *vt.*   执行
functionality   *n.*   功能性
illustrate   *vt./ vi.*   说明
instantiate   *vt.*   例示
interchangeably   *adv.*   可交换地，可替换地
interface   *n.*   接口
likewise   *adv.*   同样地

omit   *vt.*   省略
polymorphism   *n.*   多态性
precede   *vt.*   在……之前
pure   *adj.*   纯的
redefine   *vt.*   重新定义
resolve   *vi.*   解决
restrict   *vt.*   限制
scenario   *n.*   情景
sequentially   *adv.*   连续地
signature   *n.*   特征
slot   *n.*   空位置
static binding   静态绑定
subset   *n.*   子集
upcase   *n.*   向上
variable   *n.*   变量

Exercises

1. Write the output of the following program.

```cpp
class Shape{
public:
    static int number;
    double xCoord,yCoord;
    Shape(double x,double y):xCoord(x),yCoord(y)
    {
        cout << "Shape's constructor" << endl;
    }
    virtual double Area() const
    {
        cout << "Shape's area is";
        return 0.0;
    }
    ~Shape()
    {
        cout << "Shape's destructor" << endl;
    }
};
```

```cpp
class Circle:public Shape{
public:
    Circle(double x,double y):Shape(x,y)
    {
        cout << "Circle's constructor" << endl;
        number++;
    }
    virtual double Area() const
    {
        cout << "Circle's area is";
        return 3.14*xCoord*xCoord;
    }
    ~Circle()
    {
        cout << "Circle's destructor" << endl;
    }
};
class Rectangle:public Shape{
public:
    Rectangle(double x,double y):Shape(x,y)
    {
        cout << "Rectangle's constructor" << endl;
        number++;
    }
    virtual double Area() const
    {
        cout << "Rectangle's area is";
        return xCoord*yCoord;
    }
    ~Rectangle()
    {
        cout << "Rectangle's destructor" << endl;
    }
};
void fun(const Shape &sp)
{
        cout << sp.Area() << endl;
}
int Shape::number = 0;
int main()
{
    Circle c(2.0,5.0);
    Rectangle r(2.0,4.0);
    fun(c);
    fun(r);
    cout << c.number << endl;
    cout << r.number << endl;
```

```
        return 0;
}
```

2. Write the output of the following program.

```
class A{
public:
    A()
    {   t();
        cout << "i from A is " << i << endl;
    }
    void t()
    {   setI(20);    }
    virtual void setI(int m)
    {   cout << "Set a value in Class A\n ";
        i = 2 * m;   }
protected:
    int i;
};
class B: public A{
public:
    B()
    {   t();
        cout << "i from B is " << i << endl;
    }
    virtual void setI(int m)
    {   cout <<"Set a value in Class B\n ";
        i = 3 * m;   }
};
int main()
{
    A* p = new B();
    return 0;
}
```

3. Given the definition of a *Person* class as follows:

```
class Person{
public:
    Person(string n, int i):name(n), id(i){}
    virtual int Time() = 0;      //A person spends time doing something.
    virtual void print() { cout << name << id << endl; }
private:
    string name;
    int id;
};
```

Design the two classes *Student* and *Teacher*. They are derived from class *Person*.

(1) Class *Student* has the properties of *name*, *id*, *classNo* and *studyTime* per week;
Class *Teacher* has the properties of *name*, *id*, *department* and *workTime* per week.
(2) Calculate the work/student time. Calculating methods are defined as follows:
The study time of a student is the class quantity multiplied by 2 (hour) per week;
The work time of a teacher is the teaching quantity multiplied 2 (hour) per week.
(3) Display the information for classes *Student* and *Teacher*.

4. Given the definition of class *Point* as follows:

```
class Point{
public:
        Point(int xx, int yy) { x = xx; y = yy; }
        virtual void print() const { cout << x << y << endl; }
private:
        int x, y;
};
```

Design a *Circle* class with the properties of a center and a radius. The center of the circle is a point. The *Circle* class must be derived from the *Point* class. The *Circle* class can implement the operations of calculating the area and printing out the center, radius and area.

Design another class *Cylinder* with the properties of a base and a height. The base is a circle. Derive this class from the *Circle* class. The *Cylinder* class can implement the operations of calculating the volume, and printing out the base, height and volume.

Suppose the following code is in the *main* function.

```
int main()
{
    Cylinder cy(10, 20, 3, 4);
    Circle c(23, 45, 30);
    Point *p;
    p = &c;
    p->print();
    p = &cy;
    p->print();
    return 0;
}
```

5. Given the following code:

```
class Employee {
public:
    Employee(string theName, float thePayRate);
    string getName() const;
    float getPayRate() const;
    virtual float pay(float hoursWorked);
protected:
```

```
        string name;
        float payRate;
};
```

(1) Complete the definition of class *Employee*.

   Note: employee's payment = payRate * hoursWorked.

(2) Define a *Manager* class derived from class *Employee*. It has the members of *Employee* and its own member *level*.

   Note: manager's payment = payRate * level * hoursWorked.

(3) Define another class *Salesman* derived from class *Employee*. It also has the members of *Employee*, in addition to its own member *sale*.

   Note: salesman's payment = payRate * sale * hoursWorked .

(4) Test the *Manager* and *Salesman* classes by using a base class pointer.

6. Here is a class hierarchy, shown in Figure 7-5, to represent various media objects:

Given the definition of the base class *Media*:

```
class Media {
public:
        Media(){}
        virtual void print() = 0;
        virtual string id() = 0;
protected:
        string title;
};
```

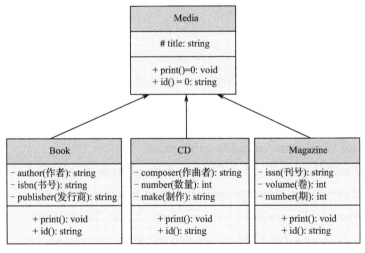

Figure 7-5　The hierarchy diagram of these media objects

(1) Explain the class relationship described in the hierarchy diagram.

(2) Define the *Book* and *Magazine* classes according to the hierarchy diagram. (Hint: The *print* function of each class displays all information of data members. The *id* function of each

class returns a string identifier that can indicate the media feature. This identifier consists of data members of each class.)

(3) Write the *main* function to test the *Book* and *Magazine* classes. (Hint: You must define a pointer to a *Media* object in the *main* function.)

7. Designs a class named *Person* and its derived classes named *Student* and *Employee*. Assume that the declaration of the *Person* class is as follows:

```
class Person{
public:
    Person(string strN, string strA): name(strN), address(strA){}
    virtual string toString() const;
private:
    string name, address;
};
```

(1) Define the *toString* function in the *Person* class. The returning value of the *toString* function is used to display the class information in the following format: Name: XXXX; Address: XXXXXXX.

(2) Define the *Student* class derived from the *Person* class. A student has a class status (freshmen, sophomore, junior, or senior). Override the *toString* function in the *Student* class.

(3) Define the *Employee* class derived from the *Person* class. An employee has an office and a salary. Override the *toString* function in the *Employee* class.

(4) Write a *top-level* function with a parameter of *Person* type and write the *main* function to test the *Student* and *Employee* classes.

8. Write a program.

(1) Abstract base class *Shape*.

(2) Classes *Triangle*, *Square* and *Circle* are derived from *Shape*.

(3) Calculate the areas and perimeters of *Triangle*, *Square* and *Circle*.

9. Design a class hierarchy as follows:

(1) A base class *Shape* with virtual function *print*.

(2) Classes *TwoDShape* and *ThreeDShape* derived from *Shape*. *TwoDShape* has virtual functions *area* and *perimeter*. *ThreeDShape* has virtual function *volume*.

(3) Classes *Triangle*, *Square* and *Circle* are derived from *TwoDShape*.

(4) Classes *Cube*, *Cuboid* and *Sphere* are derived from *ThreeDShape*.

Override the definition of the print function of each derived class. Test the hierarchy by using the *main* function and a *top-level* function with a pointer to class *Shape*.

Chapter 8 Templates

Your quote here.
—B.Stroustrup

Objectives

- To use function templates to conveniently create a group of related (overloaded) functions
- To distinguish between function templates and function template instantiation
- To use class templates to create a group of related types
- To distinguish between class templates and class template instantiation

8.1 Introduction to Templates

Many C++ programs use common data structures like *stacks*, *queues* and *lists*. A program may require a queue of customers and a queue of messages. One could easily implement a queue of customers, then take the existing code and implement a queue of messages. The program grows, and now there is a need for a queue of orders. So just take the queue of messages and convert that to a queue of orders (copy, paste, find, and replace). Need to make some changes to the queue implementation? Not a very easy task, since the code has been duplicated in many places. Reinventing source codes is not an intelligent approach in an object-oriented environment which encourages reusability. It seems to make more sense to implement a queue that can contain any arbitrary type rather than duplicating code. How does one do that? The answer is to use type parameterization, or more commonly referred to as templates.

To avoid rewriting code that would be identical except for different types. Sometimes you do not simply rely on implicit type conversion or promotion match, and you cannot stuff everything into a class hierarchy.

Thus, C++ templates allow one to implement a generic *Queue<T>* template that has a *T* type parameter. *T* can be replaced with actual types, for example,

```
Queue<Customers>
```

and C++ will generate the *Queue<Customers>* class. Changing the implementation of the *Queue* becomes relatively simple. Once the changes are implemented in the template

Queue<T>, they are immediately reflected in classes *Queue<Customers>*, *Queue<Messages>*, and *Queue<Orders>*.

Templates are very useful when implementing generic constructs like *vectors*, *stacks*, *lists*, *queues* which can be used with any arbitrary type. C++ *templates* provide a way to re-use source code as opposed to inheritance and composition that provide a way to reuse object code.

> **Generic programming** is a style of programming in which algorithms are written in terms of types to-be-specified-later. Then these types working as parameters are instantiated when specific types are needed.
>
> **Templates** provide direct support for generic programming, that is, programming using types as parameters. The C++ template mechanism allows a type to be a parameter in the definition of a class or a function.
>
> A template is a blueprint or formula for creating a generic class or a function.
>
> 泛型程序设计是一种程序设计风格，在这种程序设计中，算法是以待指定类型的形式编写的。这些类型作为参数，当需要指定类型时再实例化它们。
>
> 模板是直接支持泛型程序设计的，即程序设计使用类型作为参数。C++模板机制允许在类或函数定义中类型是一个参数。
>
> 模板是创建泛型类或函数的蓝图或准则。

C++ provides two kinds of templates: ***class templates*** and ***function templates***. Use function templates to write generic functions that can be used with arbitrary types. For example, one can write searching and sorting routines that can be used with any arbitrary type. The generic algorithms in the ***Standard Template Library*** (i.e. STL) have been implemented as function templates, and the containers have been implemented as class templates.

Inheritance and **composition** provide a way to reuse the code for **objects**. The **template** feature in C++ provides a way to reuse **source** codes.
继承和组合提供了重用对象代码的方法。C++中的**模板**特性提供了一种重用**源代码**的方法。

8.2 Function Templates

8.2.1 Definition of Function Templates

Let's imagine that we want to write a function to compare two values and indicate whether the first is less than, equal to, or greater than the second one. In practice, we would want to define several such functions, each of which could compare values of a given type. Our first attempt might be to define several overloaded functions:

```
int compare(const string &v1, const string &v2)
```

```
        {
            if (v1 < v2) return -1;
            if (v2 < v1) return 1;
            return 0;
        }
        int compare(const double &v1, const double &v2)
        {
            if (v1 < v2) return -1;
            if (v2 < v1) return 1;
            return 0;
        }
```

The two functions are nearly identical. The only difference between them is the type of their parameters. The function body is the same in each function. Having to repeat the body of the function for each type that we compare is tedious and error-prone. More importantly, we need to know in advance all the types that we might ever want to compare. This strategy cannot work if we want to be able to use the function on types that we don't know about. Therefore, we think that these *compare* functions need to be defined as a generic function.

Rather than defining a new function for each type, we can define a single ***function template***. A function template is a type-independent function that is used as a formula for generating a type-specific version of the function.

> A **function template** is a special function that can operate with generic types. In C++, the function template can be achieved using template parameters.
> 函数模板是一种使用泛型类型进行操作的特殊函数。在C++中，函数模板可以使用模板参数来实现。

This allows us to create a function template whose functionality can be adapted to more than one type or class without repeating the entire code for each type.

A template parameter is a special kind of parameter that can be used to pass a type as argument: just like regular function parameters can be used to pass values to a function, template parameters also allow to pass the types to a function. These function templates can use these parameters as if they were any other regular types.

The format for declaring a function template with type parameters is

Syntax
   template <typename identifier>
   function_declaration;
   or
   function_definition{ /*function_body*/}

where *template* and *typename* are the keywords; *identifier* is a template parameter name; *function_declaration* (or *function_definition*) includes the name, return type and parameter list of a function.

Sometimes we can define a function template in the following way:

Syntax   template <class identifier> function_declaration;

The only difference between the two prototypes is the use of either the ***typename*** keyword or the ***class***. Its use is indistinct, since both expressions have exactly the same meaning and behave exactly the same way.

For example, to define a function template that returns the greater one of two objects we could use:

```
1    template <typename T>     //you can also write: template<class T>
2    T getMax (T a, T b)
3    {
4        return (a>b? a : b);
5    }
```

Here we have created a function template with *T* as its template parameter. This template parameter represents a type that has not yet been specified, but that can be used in the function template as if it was a regular type. As you can see, the function template *getMax* returns the greater one of two parameters of this still-undefined type.

8.2.2 Function Template Instantiation

When a function template is first called for each type, the compiler creates an instantiation. Each instantiation is a version of the templated function specialized for the type. This instantiation will be called every time the function is used for the type.

Function templates can be explicitly instantiated by useing the following format:

Syntax   function_name <type> (parameters);

In the last codes, to call *getMax* to compare two integer values of type *int* we can write:

```
int x, y;
getMax <int> (x, y);
```

When the compiler encounters this call to a template instantiation, it uses the template to automatically generate a function replacing each appearance of *T* by the type passed as the actual template parameter (*int* in this case) and then calls it. This process is automatically performed by the compiler and is invisible to the programmer.

Let's make a complete example according to the definition of function template *getMax*.

Example 8-1: A function template.

```
//---------------------------------------------------------------
//File: example8_1.cpp
//This program defines a getMax function template and its instantiation.
//---------------------------------------------------------------
1    #include <iostream>
```

```
2    using namespace std;
3
4    template <typename T>
5    T getMax (T a, T b)              //function template
6    {
7        T result;
8        result = (a > b)? a : b;
9        return (result);
10   }
11   int main ()
12   {
13       int i = 5, j = 6, k;
14       long m = 10, n = 5, l;
15       k = getMax< int > (i, j);    //template instantiation
16       l = getMax< long > (m, n);   //template instantiation
17       cout << k << endl;
18       cout << l << endl;
19       return 0;
20   }
```

Result:

1	6
2	10

In this case, we have used *T* as the template parameter name not because it is shorter but in fact it is a very common template parameter name. But you can use any identifier you like.

In Example 8-1, the *getMax* function template is called twice. In Line 15, the first time with arguments of type *int*, and in Line 16, the second one with arguments of type *long*. The compiler has instantiated and then called each time the appropriate version of the function.

As you can see, the *T* type is used within the *getMax* function template even to declare new objects of that type. In Line 7, the statement

```
T result;
```

denotes that *result* will be an object of the same type as parameters *a* and *b* when the function template is instantiated with a specific type.

In this specific case where the generic *T* type is used as a parameter for *getMax* the compiler can find out automatically which data type has to instantiate without having to explicitly specify it within angle brackets (like we have done before specifying *<int>* and *<long>*). Therefore, we could have written instead:

```
int i, j;
getMax< int >(i, j)
```

Since both of *i* and *j* are of type *int*, and the compiler can automatically find out that the template parameter can only be a *int*. This implicit method produces exactly the same result.

Example 8-2: A function template without explicitly type specified.

```
//---------------------------------------------------------------
//File: example8_2.cpp
//This program defines a getMax function template and its instantiation.
//---------------------------------------------------------------
1    #include <iostream>
2    using namespace std;
3
4    template <typename T>
5    T getMax (T a, T b)
6    {
7        return (a > b? a : b);
8    }
9    int main ()
10   {
11       int i = 5, j = 6;
12       long m = 10, n = 5;
13       int k = GetMax(i, j);
14       long l = GetMax(m, n);
15       cout << k << endl;
16       cout << l << endl;
17       return 0;
18   }
```

Result:
```
1    6
2    10
```

Notice how in this case, we called function template *getMax* without explicitly specifying the type between angle-brackets (< >) in Lines 13 and 14. The compiler automatically determines what type is needed for each call.

8.2.3 Function Template with Different Parameter Types

Since the function template in Example 8-1 includes only one template parameter (typename *T*) and the function template itself accepts two parameters, both of the same *T* types, we cannot call our function template with two objects of different types as arguments. For example,

```
int i;
long l;
int k = getMin (i, l);
```

This would not be correct since the *getMax* function template expects two arguments of the same type, but in this call to it we used objects of two different types.

We can also define function templates that accept more than one type parameter, simply by specifying more template parameters between the angle brackets. For example,

```
template <typename T1, typename T2>
T1 getMin (T1 a, T2 b)
{
    return (a < b? a: (T1)b);
}
```

In this case, the *getMin* function template accepts two parameters of different types and returns a variable of the same type as the first parameter ($T1$) that is passed. For example, after that declaration, we could call the *GetMin* with the following types,

```
int i, j;
long l;
i = getMin<int, long> (j, l);
```

or simply,

```
i = getMin (j, l);
```

even though *j* and *l* have different types, because the compiler can determine the appropriate instantiation anyway.

8.2.4 Function Template Overloading

Function templates and overloading are intimately related. One can declare several function templates with the same name and even declare a combination of function templates and ordinary functions with the same name. When an overloaded function is called, overloading resolution is necessary to find the right function or template function to invoke.

For example,

```
template<typename T>
T sqrt(T);
template<typename T>
complex<T> sqrt(complex<T>);
double sqrt(double);
void f(complex<double> z)
{
    sqrt(2);        //calling function template: sqrt<int>(int)
    sqrt(2.0);      //calling ordinary function: sqrt(double)
    sqrt(z);        //calling function template: sqrt<double>(complex<double>)
}
```

A function template may be overloaded in several ways. We can provide other function templates that specify the same function name but different function parameters. A function template also can be overloaded by providing non-template functions with the same function name but different function parameters.

The compiler performs a matching process to determine what function to call when a function is invoked. First, the compiler finds all function templates that match the function named in the function call and creates specializations based on the arguments in the function call. Then, the compiler finds all the ordinary functions that match the function named in the function call. If one of the ordinary functions or function template specializations is the best match for the function call, that ordinary function or specialization is used. If an ordinary function and a specialization are equally good matches for the function call, then the ordinary function is used. Otherwise, if there are multiple matches for the function call, the compiler considers the call to be ambiguous and the compiler generates an error message.

For example,

```
template<class T>
T max(T, T);
const int s=7;
void k()
{
    max(1, 2);          //max<int>(1, 2)
    max('a', 'b');      //max<char>('a', 'b')
    max(2.7, 4.9);      //max<double>(2.7, 4.9)
    max(s, 7);          //max<int>(int(s), 7) (trivial conversion used)
    max('a', 1);        //error: ambiguous (no standard conversion)
    max(2.7, 4);        //error: ambiguous (no standard conversion)
}
```

We could resolve the two ambiguities either by explicit qualification:

```
void f()
{
    max<int>('a',1);          //max<int>(int('a'), 1)
    max<double>(2.7, 4);      //max<double>(2.7, double(4))
}
```

or by adding suitable declarations:

```
int max(int i, int j) { return max<int>(i, j); }
double max(int i, double d) { return max<double> (i, d); }
double max(double d, int i) { return max<double>(d, i); }
double max(double d1, double d2) { return max<double>(d1,d2); }
void g()
{
    max('a', 1)        //max(int('a'), 1)
    max(2.7, 4)        //max(2.7, double(4))
}
```

8.3 Class Templates

8.3.1 Definition of Class Templates

Just as we can define function templates, we can also define class templates. For example, a "*stack*" is a data structure into which we insert items at the top and retrieve those items in last-in, first-out order. It is independent of the type of the items being placed in the stack. Normally, a *Stack* class can be defined by a specified data type. If you want to create a wonderful opportunity for software reusability, you need to define a generic *Stack* class and to instantiate the class that is the type-specific versions of the generic *Stack* class. C++ provides this capability through the class template.

> A **class template** is called a parameterized type, because it requires one or more type parameters to specify how to customize a **"generic class"** template to form a class-template specialization.
> 类模板称为参数化类型，因为它们需要一个或多个类型参数来指定如何自定义"通用类"模板以形成类模板专用化。

A class template definition looks like a regular class definition, except it is prefixed by the ***template*** keyword. Now, we define a *Stack* class template in Example 8-3(a).

Example 8-3(a): A *Stack* class with parameterized types.

```
//-----------------------------------------------------------------------
//File: stack.h
//This program defines the class template of a stack structure.
//-----------------------------------------------------------------------
1   #include<iostream >
2   using namespace std;
3
4   template< typename T >
5   class Stack{
6   public:
7       Stack(int = 10);              //default constructor (Stack size 10)
8       ~Stack();                     //destructor
9       bool push(T&);                //push an element onto the Stack
10      bool pop(T&);                 //pop an element off the Stack
11
12      bool isEmpty() const;         //determine whether Stack is empty
13      bool isFull() const;          //determine whether Stack is full
14  private:
15      int size;                     //stack size
16      int top;                      //location of the top element (-1 means empty)
17      T *stackPtr;                  //pointer to the elements of the Stack
18  };
```

```
19
20   template<typename T>
21   Stack<T>::Stack(int s) : size(s > 0 ? s : 10), top(-1), stackPtr(new T[size]){}
22   template< typename T >
23   Stack<T>::~Stack()
24   {   delete[] stackPtr;   }           //delete internal space for Stack
25   template <typename T >
26   Stack<T>::push(T &pushValue)
27   {
28       if (!isFull())
29       {
30           stackPtr[++top] = pushValue;   //place item on Stack
31           return true;                    //push successful
32       }
33       return false;
34   }
35   template<typename T >
36   bool Stack<T>::pop(T &popValue)
37   {
38       if (!isEmpty())
39       {
40           popValue = stackPtr[top--];    //remove item from Stack
41           return true;                    //pop successful
42       }
43       return false;
44   }
45   template< typename T >
46   bool Stack<T>::isEmpty() const         //determine whether Stack is empty
47   {   return top == -1;   }
48   template< typename T >
49   bool Stack<T>::isFull() const          //determine whether Stack is full
50   {   return top == size - 1;   }
```

The *Stack* class template looks like a conventional class definition, except that it is preceded by the header (in Line 4)

> template< typename T >

to specify a class-template definition with type parameter *T* which acts as a placeholder for the type of the *Stack* class to be created. The programmer does not need to specifically use identifier *T* as any valid identifier can be used. The type of element to be stored on this *Stack* is mentioned generically as *T* throughout the *Stack* class header and member function definitions. Now we show how *T* becomes associated with a specific type, such as *double* or *int*. Due to the way this class template is designed, there are two constraints for non-fundamental data types used with this *Stack*. They must have a default constructor (for use in Line 21 to create the array that stores the stack elements), and they must support the assignment operator (Lines 30 and 40).

The member functions of a class template are defined as function templates. The member function definitions that appear outside the class template definition each begins with the header (in Lines 20, 22, 25, 35, 45 and 48).

> template< typename T >

Thus, each definition resembles a conventional function definition, except that the *Stack* element type is listed generically as type parameter *T*. The binary scope resolution operator is used with the class-template name *Stack<T>* to tie each member-function definition to the class template's scope. In this case, the generic class name is *Stack<T>*.

8.3.2 Class Template Instantiation

It is easy to use a class template. Create the required classes by plugging in the actual type for the type parameters. This process is commonly known as "***Instantiating a class***". Here is a sample driver class that uses class template *Stack*.

> The process of generating a class declaration from a template class and a template argument is often called **class template instantiation**.
> 从模板类和模板参数生成类声明的过程通常称为**类模板实例化**。

Now, let us consider the user program that exercises class template *Stack*. Although templates offer software-reusability benefits, remember that multiple class-template specializations are instantiated in a program (at compile time), even though the template is written only once.

Example 8-3(b): Instantiation of class template *Stack*.

```
//-----------------------------------------------------------------------
//File: example8_3.cpp
//The program instantiates class template Stack in a main function.
//-----------------------------------------------------------------------
1   #include "Stack.h"
2
3   int main()
4   {
5       Stack<double> doubleStack(5);      //size 5
6       double doubleValue = 1.1;
7       cout << "Pushing elements onto doubleStack\n";
8
9       //push 5 doubles onto a doubleStack
10      while (doubleStack.push(doubleValue))
11      {
12          cout << doubleValue << ' ';
```

```
13          doubleValue += 1.1;
14       }
15
16       cout << "\nStack is full. Cannot push " << doubleValue
17            << "\n\nPopping elements from doubleStack\n";
18
19       //pop elements from doubleStack
20       while (doubleStack.pop(doubleValue))
21          cout << doubleValue << ' ';
22       cout << "\nStack is empty. Cannot pop\n";
23
24       Stack<int> intStack;              //default size 10
25       int intValue = 1;
26       cout << "\nPushing elements onto intStack\n";
27
28       //push 10 integers onto intStack
29       while (intStack.push(intValue))
30       {
31          cout << intValue << ' ';
32          intValue++;
33       }
34
35       cout << "\nStack is full. Cannot push " << intValue
36            << "\n\nPopping elements from intStack\n";
37
38       //pop elements from intStack
39       while (intStack.pop(intValue))
40          cout << intValue << ' ';
41
42       cout << "\nStack is empty. Cannot pop" << endl;
43       return 0;
44    }
```

Result:

```
1    Pushing elements onto doubleStack
2    1.1 2.2 3.3 4.4 5.5
3    Stack is full. Cannot push 6.6
4
5    Popping elements from doubleStack
6    5.5 4.4 3.3 2.2 1.1
7    Stack is empty. Cannot pop
8
9    Pushing elements onto intStack
10   1 2 3 4 5 6 7 8 9 10
11   Stack is full. Cannot push 11
12
13   Popping elements from intStack
```

```
14    10 9 8 7 6 5 4 3 2 1
15    Stack is empty. Cannot pop
```

The program begins by instantiating an object *doubleStack* of size 5 in Line 5. This object is declared to be of class *Stack< double >* (pronounced "Stack of double"). The compiler associates type *double* with type parameter *T* in the class template to produce the source code for a *Stack* class of type *double*.

When *doubleStack* is instantiated as type *Stack<double>*, the specialization of the *Stack* constructor template uses the **new** operator to create an array of elements of type *double* to represent the stack (in Line 21 of Example 8-3(a)). The statement in the *Stack* class template definition

```
stackPtr = new T[ size ];
```

is generated by the compiler in the class template specialization *Stack< double >* as

```
stackPtr = new double[ size ];
```

Line 10 invokes the *push* function to place the double values 1.1, 2.2, 3.3, 4.4 and 5.5 onto *doubleStack*. The while loop terminates when the program attempts to push the sixth value onto *doubleStack* (which is full, because it holds a maximum of five elements). Note that the *push* function returns false when it is unable to push a value onto the stack. Class *Stack* provides the *isFull* function, which the programmer can use to determine whether the stack is full before attempting a push operation. This would avoid the potential error of pushing onto a full stack.

Lines 20 and 21 invoke the *pop* function in a while loop to remove the five values from the stack, that the values do pop off in last-in, first-out order). When the program attempts to pop the sixth value, the *doubleStack* is empty, so the *pop* loop terminates. Line 24 instantiates an integer stack class *intStack* with the declaration

```
Stack< int > intStack;
```

Since no size is specified, the size defaults to 10 as specified in the default constructor (in Line 7 of Example 8-3(a)). Lines 29 through 33 loop and invoke *push* to place values onto *intStack* until it is full, then Lines 39 and 40 loop and invoke *pop* to remove values from *intStack* until it is empty. Once again, notice in the output that the values pop off in last-in, first-out order.

8.4 Non-Type Parameters for Templates

Besides the template arguments that are preceded by the **typename** keywords, which represent types, templates can also have regular typed parameters, like those found in functions. As an example, take a look at this class template that is used to contain sequences of elements.

Example 8-4: Non-type parameters for templates.

```
//--------------------------------------------------------------------
//File: example8_4.cpp
//This program defines a non-type parameter for a class template.
//--------------------------------------------------------------------
1   #include <iostream>
2   using namespace std;
3
4   template <typename T, int N>
5   class mysequence {
6       T memblock [N];
7   public:
8       void setMember (int x, T value);
9       T getMember (int x);
10  };
11
12  template <typename T, int N>
13  void mysequence<T, N>::setMember (int x, T value) {
14      memblock[x] = value;
15  }
16
17  template <typename T, int N>
18  T mysequence<T, N>::getMember (int x) {
19      return memblock[x];
20  }
21
22  int main ()
23  {
24      mysequence <int, 5> myints;
25      mysequence <double, 5> myfloats;
26      myints.setMember (0, 100);
27      myfloats.setMember (3, 3.1416);
28      cout << myints.getMember(0) << '\n';
29      cout << myfloats.getMember(3) << '\n';
30      return 0;
31  }
```

Result:

| 1 | 100 |
| 2 | 3.1416 |

In this example, integer N is a non-type template parameter. When the objects of the template are created in Lines 24 and 25, an integer 5 is directly passed into N. It is also possible to set default values or types for class template parameters. For example, if the previous class template definition had been:

```
template <typename T = char, int N=10>
class mysequence {...};
```

We could create objects using the default template parameters by declaring:

```
mysequence<> myseq;
```

which would be equivalent to:

```
mysequence<char,10> myseq;
```

8.5 Derivation and Class Templates

Class templates can inherit or be inherited from. For many purposes, there is nothing significantly different between the template and non-template scenarios. However, there is one important subtlety when deriving a class template from a base class referred to by a dependent name. Let's first look at the somewhat simpler case of nondependent base classes.

In a class template, a nondependent base class is one with a complete type that can be determined without knowing the template arguments. In other words, the name of this base is denoted by using a nondependent name. For example:

```
1   template<typename T>
2   class Base {
3   public:
4       int basefield;
5       typedef int T;
6   };
7
8   class D1: public Base<Base<void> > {    //not a template case really
9   public:
10      void f() { basefield = 3; }          //usual access to inherited member
11  };
12
13  template<typename T>
14  class D2 : public Base<double> {         //nondependent base
15  public:
16      void f() { basefield = 7; }          //usual access to inherited member
17      T strange;                           //T is Base<double>::T, not the template parameter!
18  };
```

Nondependent bases in templates behave very much like bases in ordinary nontemplate classes, but there is a slightly unfortunate surprise: When an unqualified name is looked up in the templated derivation, the nondependent bases are considered before the list of template parameters. This means that in the previous example, the *strange* member of the *D2* class template always has the *T* type corresponding to *Base<double>::T* (in other words, *int*). For example, the following function is not valid in C++ (assuming the declarations above):

```
void g (D2<int*>& d2, int* p)
{
    d2.strange = p;         //error: type mismatch!
}
```

However, if we test the class declaration above in the following program, it goes well.

```
1  int main()
2  {
3      D1 d1;
4      d1.f();
5      cout << d1.basefield << endl;
6      D2<int> d2;
7      d2.f();
8      cout << d2.basefield << endl;
9      d2.strange = 8;
10     cout << d2.strange << endl;
11     return 0;
12 }
```

This is counterintuitive and requires the writer of the derived template to be aware of names in the nondependent bases from which it derives—even when that derivation is indirect or the names are private. It would probably have been preferable to place template parameters in the scope of the entity they "templatize".

Templates and Inheritance

Templates and inheritance relate in several ways:
(1) A class template can be derived from a class-template specialization.
(2) A class template can be derived from a non-template class.

(3) A class-template specialization can be derived from a class-template specialization.

(4) A non-template class can be derived from a class-template specialization.

模板与继承

模板和继承有几种方式：

（1）类模板可以从类模板特例化中派生。

（2）类模板可以从非模板类派生。

（3）类模板特例化可以从类模板特例化派生。

（4）非模板类可以从类模板特例化派生。

Think These Over

1. What is a template? Why are templates used?

2. What is a function template?

3. What is the meaning of template instantiation?

4. What is a class template?

8.6 Case Study: A Vector Class Template

You can use an array to store a collection of data such as int, float or user-defined type. There is a serious limitation—the array size is fixed when the array is declared. C++ provides a ***vector*** class, which is more flexible than arrays. You can use a ***vector*** object just like an array. The vector size can increase automatically while an element is stored in the vector. Now we define a *Vector* class template and initialize it by using the *Student* class objects and *float* values.

Example 8-5: A *Vector* class with parameterized types.

```
//-----------------------------------------------------------------
//File: example8_5.cpp
//This program defines the class template of a vector structure and initializes by using the Student
//class objects and float values.
//-----------------------------------------------------------------
1   #include <iostream>
2   #include <cstring>
3
4   using std::cout;
5   using std::cin;
6   using std::endl;
7
8   //definition of the Student class
9   class Student {
10    public:
11      Student(char *strname = "", char *id = " ", int g = 0)
12      {
13          name = new char[strlen(strname) + 1];
14          strcpy(name, strname);
15          ID = new char[strlen(id) + 1];
16          strcpy(ID, id);
17          grade = g;
18      }
19      Student(const Student& s)
20      {
21          name = new char[strlen(s.name) + 1];
22          strcpy(name, s.name);
```

```
23        ID = new char[strlen(s.ID) + 1];
24        strcpy(ID, s.ID);
25        grade = s.grade;
26     }
27     Student& operator =(const Student &s)
28     {
29        delete[] name;
30        name = new char[strlen(s.name) + 1];
31        strcpy(name, s.name);
32
33        delete[] ID;
34        ID = new char[strlen(s.ID) + 1];
35        strcpy(ID, s.ID);
36
37        grade = s.grade;
38        return *this;
39     }
40     operator int()
41     { return -1000; }
42     void print() const
43     { cout << "Student: " << ID << " " << name << "; grade: " << grade << endl; }
44     ~Student()
45     {
46        delete[] name;
47        delete[] ID;
48     }
49  private:
50     char *name, *ID;
51     int grade;
52  };
53
54  //definition of a Vector class template
55  template<typename T>
56  class Vector {
57  public:
58     Vector() :size(0), values(0), space(0){}
59     Vector(int len)                          //constructor
60     {
61        size = 0;
62        space = len;
63        values = new T[len];
64     }
65     ~Vector()                                //destructor
66     { delete[]values; }
67     Vector(const Vector& v)                  //copy constructor
68     {
69        size = v.size;
```

```cpp
70            values = new T[size];
71            for (int i = 0; i <size; ++i)
72                values[i] = v.values[i];
73            space = v.space;
74        }
75
76        Vector& operator = (const Vector& v);        //overloaded operator =
77        T& operator[] (int index)
78        { return values[index]; }
79        T at(int index);
80
81        template<typename T>
82        friend std::ostream& operator << (std::ostream&, Vector<T>&); //output a Vector
83
84        void push_back(T d);
85        void pop_back();
86
87        int getSize() const { return size; }          //get the size of the Vector
88        void reserve(int newalloc);
89        void resize(int newsize);
90        int capacity() const { return space; }
91
92        bool empty() const { return size == 0; }
93        void clear() { size = 0; }
94   private:
95        T *values;                                    //elements of the Vector
96        int size;                                     //size of the Vector
97        int space;
98   };
99
100  template <typename T>
101  Vector<T>& Vector<T>::operator=(const Vector<T>& v)
102  {
103       if (&v != this)
104       {
105           delete[] values;
106           values = new T[v.size];
107           for (int i = 0; i < v.size; i++)
108               values[i] = v.values[i];
109           space = v.space;
110       }
111       return *this;
112  }
113  template <typename T>
114  void Vector<T>::push_back(T d)
115  {
116       if (space == 0) reserve(8);
```

```cpp
117        else if (space == size) reserve(2 * space);
118        values[size] = d;
119        ++size;
120    }
121    template <typename T>
122    void Vector<T>::reserve(int newalloc)
123    {
124        if (newalloc <= space) return;
125        T *p = new T[newalloc];
126        for (int i = 0; i < size; i++)
127            p[i] = values[i];
128        delete[] values;
129        values = p;
130        space = newalloc;
131    }
132    template <typename T>
133    void Vector<T>::resize(int newsize)
134    {
135        reserve(newsize);
136        for (int i = size; i <newsize; ++i) values[i] = 0;
137        size = newsize;
138    }
139    template <typename T>
140    void Vector<T>::pop_back()
141    {
142        if (empty())
143            cout << "Vector is empty.\n";
144        else
145            --size;
146    }
147
148    template <typename T>
149    bool Vector<T>::empty() const
150    { return size == 0; }
151
152    template <typename T>
153    T Vector<T>::at(int index)
154    {
155        if (index >= size)
156        {
157            cout << "Beyond boundary\n";
158            return values[index];
159        }
160        else
161            return values[index];
162    }
163
```

```cpp
164  template <typename T>
165  std::ostream& operator <<(std::ostream& os, Vector<T>& v)
166  {
167      for (int i = 0; i < v.getSize(); ++i)
168          os << v.values[i] << " ";
169      return os;
170  }
171
172  int main()
173  {
174      int i;
175      Vector<Student> sVector(3);
176      cout << "sVector size: " << sVector.getSize() << " sVector' space: "
177           << sVector.capacity() << endl;
178
179      char na[512], id[512];
180      int gr;
181      for (i = 0; i < sVector. capacity(); i++)
182      {
183          cin >> na >> id >> gr;
184          Student st(na, id, gr);
185          sVector.push_back(st);
186      }
187      for (i = 0; i < sVector.getSize(); i++)
188          sVector.at(i).print();
189      cout << "Appends elements at the back of the vector\n";
190      cout << "Remove the last element from the vector\n";
191      sVector.pop_back();
192      cout << "sVector size is " << sVector.getSize() << " sVector' space is "
193           << sVector.capacity() << endl;
194
195      Vector<float> fVector(4);   //declare a template class with *float* data type
196
197      for (i = 0; i < fVector.capacity(); i++)
198          fVector.push_back(float(i * 3 + 0.5));
199      cout << "fVector=" << fVector << endl;
200      cout << "Remove the last element from the vector\n";
201      fVector.pop_back();
202      cout << "fVector size is " << fVector.getSize() << " sVector' space is "
203           << fVector.capacity() << endl;
204      cout << fVector.at(2) << endl;
205      fVector.clear();
206      cout << "fVector size is " << fVector.getSize() << " fVector' space is "
207           << fVector.capacity() << endl;
208      cout << "fVector = " << fVector << endl;
209      return 0;
210  }
```

Input:

wang 13001201 78
gao 13001202 90
zhang 13001203 67

Result:

1 sVector size: 0 sVector' space: 3
2 Student: 13001201 wang; grade: 78
3 Student: 13001202 gao; grade: 90
4 Student: 13001203 zhang; grade: 67
5 Appends elements at the back of the vector
6 Remove the last element from the vector
7 sVector size is 2 sVector' space is 3
8 fVector=0.5 3.5 6.5 9.5
9 Remove the last element from the vector
10 fVector size is 3 sVector' space is 4
11 6.5
12 fVector size is 0 fVector' space is 4
13 fVector =

Word Tips

actual *adj.* 实际的，目前的
associate *vt.* 使联合
automatically *adv.* 自动地
angle *n.* 角度
appropriate *adj.* 适当的
arbitrary *adj.* 任意的，独裁的
container *n.* 容器
conveniently *adv.* 方便地，合宜地
double-stack *n.* 双栈结构
error-prone *adj.* 易错的
encounter *vt./vi.* 遭遇，遇到
first-out 先出
generic *adj.* 普通的; *n.* 泛型
generically *adv.* 一般地
intelligent *adj.* 聪明的，智能的
indistinct *adj.* 不清楚的，模糊的
instantiation *n.* 实例
identical *adj.* 同一的，同样的
last-in 后进

non-template 非模板
placeholder *n.* 占位符
plug *n.* 插头
paste *vt.* 粘贴
parameterization *n.* 参数化
regular *adj.* 规则的
retrieve *vt.* 检索，重新得到
reuse *vt.* 重用
relate as 与……有关
reinvent *vt.* 更改，重做
reusability *n.* 复用性
stack *n.* 栈
software-reusability 软件重用性
strategy *n.* 策略
specialization *n.* 专门化，特例化
template *n.* 模板
tedious *adj.* 单调的，乏味的
version *n.* 版本
vectors *n.* 向量

Exercises

1. Find the error(s) in each of the following lines and explain how to correct it.

```
template<class T>
T min(T, T);
const int s=7;
void k( )
{
    min(5,2);
    min('a', 'b');
    min('a',1);
    min(s,7);
    min(2.7,4);
}
```

2. Define a *Swap* function template that implements exchange of two numbers. Implement the template by using types *int* and *double*.

3. Define a *getMax* function template with finding the maximum value among 10 numbers. Implement it by using types *int* and *float*.

4. Write a function template to compare the values of two values. If they are equal, return 1; otherwise, return 0. Instantiate it by using types *double* and *string*.

5. Define a class template *array* that produces a bounds-checked array. Instantiate it by using types *int* and *float*.

6. Define a *Stack* class template. Implement the template by using *int* and *char*.

7. Given the following program:

```
template <typename T>
class Vector {
public:
    Vector(int len);
    ~Vector();
    Vector (const Vector<T>& v);
    T operator ()(int i) const;
    T& operator()(int i);
    int getsize()                                       //get the size of the Vector
    Vector<T>& operator= (const Vector<T>& v);          //overloaded operator =
    friend ostream& operator << (ostream&, Vector<T>&); //output a vector
private:
    T *values;                                          //the element of the Vector
    int size;                                           //size of the Vector
};
```

Write out the definition of the *Vector* class template.

8. Write a class template to search an element in an ordered array by using *dichotomy*.

References

[1] STROUSTRUP B. The C++ Programming Language[M]. Special Edition. Pearson Press，2001.

[2] ECKEL B. Thinking in C++[M]. Prentice Hall Inc.，1995.

[3] MALIK D S. C++ Programming from Problem Analysis to Program Design[M]. 3th ed. Thomson Course Technology，2007.

[4] DEITEL P J，DEITEL H M. C++ How to Program[M]. 6th ed. Prentice Hall Inc.，2008.

[5] LIPPMAN S B，LAJOIE J，BARBARA E M. C++ Primer[M]. 5th ed. 王刚，杨巨峰，译. 北京：电子工业出版社，2013.

[6] STROUSTRUP B. C++程序设计原理与实践[M]. 王刚，等译. 北京：机械工业出版社，2010.

[7] SAVITCH W. Problem Solving with C++：The Object of Programming[M]. 3th ed. 周靖，译. 北京：清华大学出版社，2005.

[8] http://www.cplusplus.com/doc/tutorial/.

反侵权盗版声明

电子工业出版社依法对本作品享有专有出版权。任何未经权利人书面许可,复制、销售或通过信息网络传播本作品的行为,歪曲、篡改、剽窃本作品的行为,均违反《中华人民共和国著作权法》,其行为人应承担相应的民事责任和行政责任,构成犯罪的,将被依法追究刑事责任。

为了维护市场秩序,保护权利人的合法权益,我社将依法查处和打击侵权盗版的单位和个人。欢迎社会各界人士积极举报侵权盗版行为,本社将奖励举报有功人员,并保证举报人的信息不被泄露。

举报电话:(010)88254396;(010)88258888
传　　真:(010)88254397
E-mail:　dbqq@phei.com.cn
通信地址:北京市海淀区万寿路 173 信箱
　　　　　电子工业出版社总编办公室
邮　　编:100036